The Earth's Core

International Geophysics Series

EDITORS

J. VAN MIEGHEM

Royal Belgian Meteorological Institute
Uccle, Belgium

ANTON L. HALES

Australian National University
Canberra, A.C.T., Australia

The Earth's Core

J. A. JACOBS

Department of Geodesy and Geophysics
University of Cambridge, England

1975

ACADEMIC PRESS

London   New York   San Francisco

A Subsidiary of Harcourt Brace Jovanovich, Publishers

ACADEMIC PRESS INC. (LONDON) LTD
24/28 Oval Road
London NW1

United States Edition published by
ACADEMIC PRESS INC.
111 Fifth Avenue
New York, New York 10003

Library of Congress Catalog Card Number: 75–19654
ISBN: 0–12–378950–8

PRINTED IN GREAT BRITAIN BY
ADLARD AND SON LTD., BARTHOLOMEW PRESS, DORKING

Preface

The innermost regions of the Earth are inaccessible to man and no direct measurements of any of its physical properties can be made. Much attention has been given in the last few years to the "inverse" problem in geophysics—that of determining some physical parameter from a set of observations made at the surface of the Earth. Our knowledge of the core of the Earth comes from many different fields, of which seismology and geomagnetism are the most important and a discussion of these disciplines forms a large part of this book.

The role of the Earth's core is essential to our understanding of many geophysical phenomena—it is the seat of the Earth's magnetic field and coupling between the core and mantle is responsible for some of the variations in the length of the day. The advent of satellites and spacecraft to the moon and terrestrial planets has given an added interest to the internal properties of these bodies—measurements of their magnetic field has already given us some clues on their constitution and possible cores. The question of cores in the other planets is discussed in the last chapter.

J. A. Jacobs

July 1975

To Peggy

Contents

1. General Physical Properties of the Earth

1.1. Introduction

This book is concerned with the Earth's core—the innermost regions of the Earth. No direct measurements can be made of any of its physical properties and in some ways we know more about the distant stars than we do about the deep interior of the Earth. However the core plays a key role in many geophysical studies—amongst other things it is believed that the Earth's magnetic field is a result of motions in the fluid outer core (Section 4.2).

The core, whose radius is just over one half that of the Earth, consists of an outer core which is fluid and an inner core which is most probably solid—there may also be one or more transition zones between these two regions. The radius of the inner core is about one-fifth that of the Earth—thus, although its linear dimensions are quite large, its volume is only about 0·007 that of the whole Earth.

Geophysics is an observational science and our objective is to try and explain our observations with theories based on sound scientific principles. To do this we set up models which approximate conditions in the real Earth and try and solve the problem for this idealized situation. A whole series of models of ever increasing sophistication can be considered, but it must not be forgotten that they are only models and that the real Earth may behave quite differently. The assumptions and limitations of any model must be clearly stated and always borne in mind when assessing the success of any theory.

The mathematical equations that describe most physical phenomena are often non-linear and extremely complex so that analytical solutions are in general not possible. In some cases order of magnitude arguments are often used to draw some preliminary conclusions, but one must be very wary of

1

much of this type of "arithmetic", since it is all too easy to be misled by "geophysical numerology" (Jacobs, 1970). In such cases intuition is a notoriously bad guide. Moreover order of magnitude calculations can often lead to contradictions resulting from over simplification of the original equations.

Our knowledge of the deep interior of the Earth is pieced together from information obtained from a number of different disciplines—physics, chemistry, astronomy and geology. A fundamental problem in geophysics is the determination of some physical parameter (e.g. density) from a set of observations made at the surface of the Earth. This "inverse" problem as it is called has received much attention during the last few years, particularly in the U.S.A. and the U.S.S.R., and will be discussed in some detail in Section 1.5. It has been formulated using sophisticated mathematics, and with the aid of modern computers it is possible to obtain information about the uncertainty and non-uniqueness of Earth models.

At the XV General Assembly of the International Union of Geodesy and Geophysics held in Moscow in August 1971, a committee was set up to advise on a Standard Earth Model (SEM) covering the distributions of various physical properties of the Earth's interior. A necessary requirement of a reference Earth model is that it must fit data on the Earth's mean radius R, mass M, and z where $I = zMR^2$ and I is the Earth's mean moment of inertia. Account must also be taken of data derived from records of seismic body waves, seismic surface waves, and free Earth oscillations. In addition there is a large body of evidence from other sources, including data on Earth tides, finite-strain and solid-state theory, laboratory experiments on rocks (including shock-wave experiments at pressures up to 4 million atmospheres), and evidence from other disciplines such as planetary physics, geology and geochemistry.

The SEM committee set up a number of sub-committees to examine particular aspects of the SEM project. At a meeting in Lima in August 1973, it was decided to publish some of the interim reports prepared by the sub-committees and these have now appeared in *Physics of the Earth and Planetary Interiors* (**9**, 1–44, 1974).

1.2. Travel-Time and Velocity-Depth Curves

The major source of our information about the Earth's interior comes from the field of seismology. Following an earthquake, elastic waves travel throughout the Earth and may be observed at the large number of seismological stations distributed across the world. The theory of the propagation of elastic waves may be found in any standard text on seismology (see e.g. Bullen, 1963) and will not be developed here. There are two distinct types of elastic waves—P waves and S waves. A P wave is a condensation-rarefaction wave involving change of volume. Motion of the medium is longitudinal so that there is no polarization of a P wave. An S wave is a shear wave in which there is distortion without change of volume. S waves are transverse waves and thus exhibit polarization. In addition surface waves are set up: these have yielded valuable information about the crustal layers of the Earth. Recently long period surface waves have been used to obtain additional information about the deeper parts of the Earth.

It can be shown that the velocities of P and S waves are given by

$$V_P = \sqrt{\frac{k_s + \frac{4}{3}\mu}{\rho}} \qquad (1.1)$$

and

$$V_S = \sqrt{\frac{\mu}{\rho}} \qquad (1.2)$$

where k_s is the bulk modulus or adiabatic incompressibility, μ the modulus of rigidity and ρ the density. Thus V_P and V_S depend only on the elastic parameters and density of the medium. In particular if the rigidity is zero, $V_S = 0$ i.e. shear waves cannot be transmitted through a liquid. It follows from Eqns. (1.1) and (1.2) that

$$\phi = \frac{k_s}{\rho} = V_P^2 - \frac{4}{3}V_S^2. \qquad (1.3)$$

When an elastic wave meets a sharp boundary between two media of different properties, part of it will be reflected and part refracted, and laws of reflection and refraction analogous to those of geometrical optics apply. The case of elastic waves is more complicated, however, since waves of both P and S type may be reflected and refracted. The theory is based on Fermat's principle according to which an elastic wave takes the quickest path between any two points. This does not imply that there is only one path—there may be a number of alternative paths—but each path must involve a minimum transit time relative to small deviations in the path.

Figure 1.1 illustrates some of the many possible reflections and refractions of elastic waves at discontinuities within the Earth. The figure also illustrates the terminology used to designate some of the different phases. A P wave reflected from the Earth's surface can give rise to both a P wave and an S wave (called PP and PS waves respectively). Likewise an S wave reflected from the surface can give both P and S waves (called SP and SS waves respectively). The letters c and i denote reflections from the outer and inner core boundaries respectively, and the letter h reflection from the surface of the F shell (the transition layer between the inner and outer core). The letters K and I are used for P waves in the outer and inner core. Thus SKP is an S wave that has been refracted into the outer core (necessarily as a P wave) and refracted back into the mantle as a *P* wave. *S* waves have never been observed in the outer core which is thus considered to be liquid. On the other hand, it has often been conjectured that the inner core is solid and the symbol *J* was proposed for paths of *S* waves (if they exist) in the inner core. It was not until 1972 that Julian *et al.* claimed to have identified the phase *PKJKP* on seismograms— confirmation of their work would establish the existence of rigidity in the inner core directly. They obtained a value of $V_s \simeq 2 \cdot 95$ km/sec in the inner

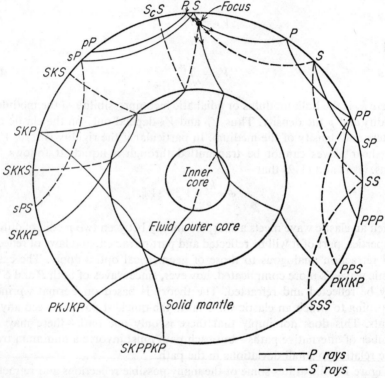

Fig. 1.1. Representative seismic rays through the Earth. (After Bullen, 1954.)

core which is difficult to reconcile with that of $\simeq 3 \cdot 6$ km/sec obtained by Gilbert *et al.* (1973) from an analysis of free Earth oscillation data (see Section 1.3). A value of $\simeq 3 \cdot 6$ km/sec also follows from Bullen's (k,p) hypothesis (see Section 5.3). There is a possibility that Julian *et al.* observed the phase *SKJKP*, rather than *PKJKP*—Doornbos (1974) has suggested that the phase *PKJKP* is too small to be observed. Multiple (*n*) internal reflections from the mantle core boundary (MCB) into the outer core are indicated by *nK*. Thus *P7KP* is a *P* wave that has been refracted into the outer core, suffering 7 internal reflections in the core before being refracted back into the mantle as a *P* wave (see Fig. 1.2).

The times of arrival of the different seismic waves may be determined from the records at a number of stations so that it is possible to construct travel-time curves, i.e. plots of arrival times *T* against distance Δ (measured in

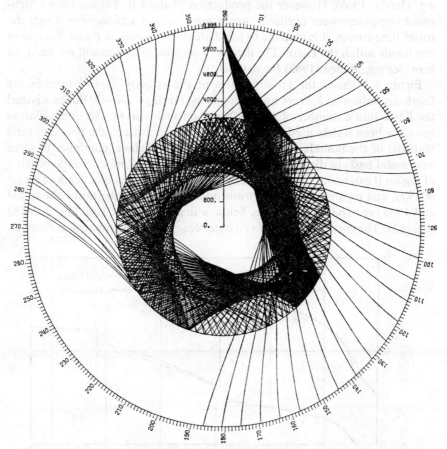

Fig. 1.2. Diagram showing seven internal reflections from the MCB into the outer core (*P7KP*). After C. H. Chapman.

degrees along the surface of the Earth) between the source and the seismic observatory. The construction of travel-time curves has had a long history of successive approximation and increasing accuracy. Revision of the early travel-time curves was undertaken by Jeffreys in 1931 (see Jeffreys, 1936, 1961 for complete details) using a least-squares technique. In collaboration with Bullen he produced the first J.B. Tables in 1935. Substantial refinements were incorporated in a new set of J.B. Tables first published in 1940, these gave travel-times, not only of *P* and *S* waves, but also of reflected and refracted waves. Details of the production of these tables and of the early history of the subject may be found in Bullen (1963). Improvements in the records and the use of large artificial explosions have led to corrections to the J.B. Tables (see e.g. Herrin, 1968). However the production of the J.B. Tables, before high-speed computers were available, was a monumental achievement. From the travel-time curves, it is possible to calculate the velocities of *P* and *S* waves at any depth within the Earth. The details of such an inversion will not be given here (see e.g. Bullen (1963) for details).

Figure 1.3 shows the gross features of the velocity–depth curves in the Earth according to Jeffreys and Gutenberg. During 1940–42, Bullen divided the Earth into a number of regions based on such curves. His nomenclature has since been widely used, and, in spite of uncertainties in the positions (and realities) of the boundaries between the different regions, continues to serve as a useful basis in discussing the Earth's interior. The upper mantle consists of region B extending from the base of the crust (region A) to a depth of about 400 km, and region C which is a transition zone between depths of about 400 and 1000 km. The lower mantle, below a depth of about 1000 km, is called region D. The core is divided into an outer core E, a transition region F and an

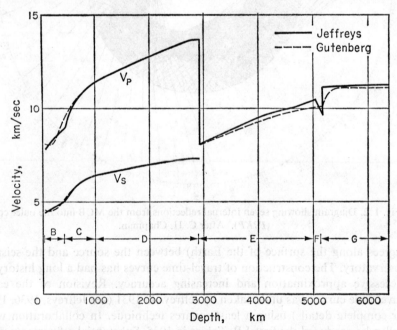

Fig. 1.3. Seismic velocities of *P* and *S* waves as a function of depth. (After Birch, 1952.)

inner core G. In recent years much finer detail has been obtained, particularly in the upper mantle (regions B and C, see Fig. 1.4), and in the vicinity of the boundary of the inner core (region F). In the following discussion of the physical properties of the Earth, attention will be confined to the core and very deep mantle: no account will be given of the crust, the low velocity layer in the upper mantle nor of possible phase changes in region C.

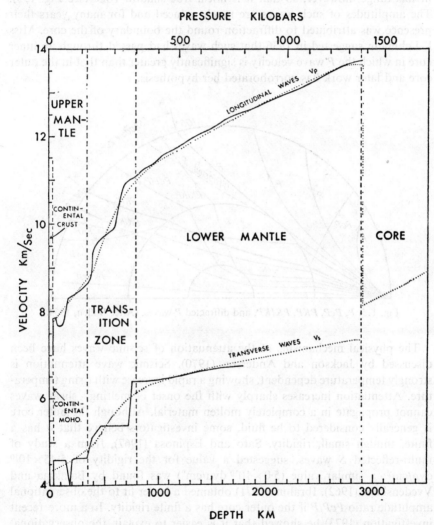

Fig. 1.4. Seismic velocity distribution in the mantle for *P* waves, solid line (Johnson, 1967, 1969); *S* waves, solid line (Nuttli, 1969); broken lines *P* and *S* waves (Jeffreys, 1937, 1939c). (After Ringwood, 1972.)

The inversion methods of obtaining velocity–depth curves from travel-time curves break down if there is a region in the Earth in which the velocity decreases with depth. In such a case there will be a range of Δ in which arrivals from an earthquake are not observed (a shadow zone). There is an abrupt, discontinuous drop in the velocity of *P* waves across the MCB with a corresponding shadow zone in the range 105° < Δ < 143°. Some waves are recorded in this range, however, so that it is not a true shadow zone (see Fig. 1.5). The amplitudes of such waves are much reduced and for many years their presence was attributed to diffraction round the boundary of the core. Miss I. Lehmann suggested in 1936 that such waves had passed through an inner core in which the *P* wave velocity is significantly greater than that in the outer core and later work has corroborated her hypothesis.

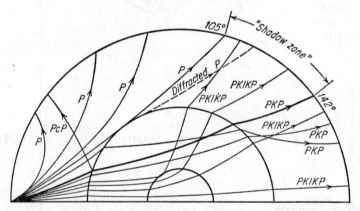

Fig. 1.5. *P*, *PcP*, *PKP*, *PKIKP*, and diffracted *P* waves. (After Bullen, 1954.)

The physical mechanisms of the attenuation of seismic waves have been discussed by Jackson and Anderson (1970). Seismic wave attenuation is strongly temperature dependent, showing a rapid increase with rising temperature. Attenuation increases sharply with the onset of melting—shear waves cannot propagate in a completely molten material. Although the outer core is generally considered to be fluid, some investigators believe that it has a finite, though small, rigidity. Sato and Espinosa (1967), from a study of multi-reflected *S* waves, suggested a value for the rigidity of $5 \cdot 45 \times 10^{10}$ dyn/cm^2: a similar value ($5 \cdot 0 \times 10^{10}$ dyn/cm^2) was found by Balakina and Vvedenskaya (1962). Ibrahim (1971) obtained a better fit to the observational amplitude ratio *PcP/P* if the outer core has a finite rigidity. In a more recent investigation (1973) he showed that it is easier to explain the observational data (mainly from deep focus earthquakes) if two to four low velocity, high density layers are imbedded between the mantle and core. Ibrahim constructed a number of different models using different spectral ratios for different phases—Table 1.1 gives the results for one particular model. All models indicate that the outer core has a finite rigidity, with an upper bound for the shear wave velocity of 1·4 km/sec.

Buchbinder and Poupinet (1973) have criticized Ibrahim's results and claim that the layered system he proposed is not likely to be found from the data he used (short period phases sampled at their maximum amplitude), even if such a layered system existed. Buchbinder and Poupinet maintain that PcP/P amplitude ratios cannot be used to study the structure of the MCB. However, from a study of their wave forms (following two large nuclear explosions), they suggest that the MCB may be approximated by a thin high-impedance liquid layer several kilometres thick embedded between the mantle and the core—the data do not permit exact determination of the model parameters, and the structure of the MCB may be radially asymmetric.

Table 1.1. Structural model for the core–mantle boundary used in P/PcP calculations. (After Ibrahim 1973.)

Region	P-wave velocity (km/sec)	S-wave velocity (km/sec)	Density (g/cm³)	Layer thickness (km)
Lower mantle	13·50	6·95	5·65	
Imbedded layer no. 1	11·60	6·10	5·67	12·00
Imbedded layer no. 2	10·20	5·20	6·65	8·00
Outer core	8·15	1·40	9·40	

Values of Q^* in the outer core in Ibrahim's models range from 100 to 1000 which may indicate that it is chemically zoned. These values should be compared with those of Sacks (1971) whose values range from 3000 to 10,000 and those of Adams (1972) who obtained a lower bound of about 2200 for Q in the outer core. It must be pointed out, however, that part of these differences in Q-values may be due to differences in the period ranges upon which the estimates were based. Qamar and Eisenberg (1974) have investigated attenuation in the core using spectral ratios of seismic core waves. For

---

* Departures from ideal elasticity in the Earth can be expressed in terms of the reciprocal of the dimensionless parameter Q (Knopoff, 1964), defined by the equation

$$\frac{2\pi}{Q} = \frac{\Delta E}{E}$$

where ΔE is the energy dissipated per cycle and E is the peak energy stored. The mechanical Q defined above is mathematically equivalent to the Q of an oscillatory electrical circuit. A region with low Q values thus implies an extremely dissipative zone.

seismic waves of frequency $f \geqslant 1$ Hz they found a high value of $Q(\geqslant 5000)$ in the outer core. This was based on the fact that $PnKP$ phases were observed for large values of n (at least $n = 7$): the $P7KP/P4KP$ ratios suggest a Q in the range 4000–8000, while the average slope of the $P7KP/P4KP$ spectral ratio yields $Q \simeq 10,000$. However, the results of Qamar and Eisenberg, which were carried out over a very narrow frequency band near 1 Hz, are in disagreement with the value of $Q = 100$–300 determined by Suzuki and Sato (1970) for $0.04 < f < 0.15$ Hz. For the outer 450 km of the inner core, Qamar and Eisenberg, using the $PKIKP/PKP$ ratio, obtained a value of Q in the range 120–1400, more than an order of magnitude less than that in the outer core. The difference presumably reflects the solidity of the inner core.

Doornbos (1974) has studied the anelasticity of the inner core using spectral ratios of short-period core phases with common source and receiver and with nearly the same ray paths in the mantle. For 1 Hz compressional waves, he found that Q rises from a value of about 200 near the inner core—outer core boundary to about 600 at a depth of 400 km inside the inner core. Below 450 km from the boundary, the Q structure cannot be determined with any precision but is probably less than 2000. From observations of high Q normal modes with a high energy concentration in the inner core, Doornbos found a suggestion that over a wider frequency range, Q is frequency dependent. A frequency dependent Q with low Q values around 1 Hz is compatible with partial melting in the inner core. He suggested that at the boundary of the inner core—outer core, the temperature is close to the melting point and that in the inner core the temperature gradient only slightly exceeds the melting point gradient (see also Section 3.5).

Our present knowledge of seismic velocities in the core comes chiefly from the interpretation of travel-times of short period P waves. There is a wide variation in the estimated velocities of P waves just below the MCB with likely values ranging from 7.9 km/sec (Hales and Roberts, 1971) to 8.26 km/sec (Randall, 1970). Inversion studies favour a value fairly close to Jeffreys' (1939d) value of 8.10 km/sec. At depths in the outer core greater than about 200 km, most velocity models converge to values that are generally higher than those of the Jeffreys (1939d) model, although within 0.1 km/sec of it. A detailed discussion of the results of different authors has been given by Engdahl (1968). Nearly all authors of travel-time studies agree that there is a gradual increase in P velocity in the outer core down to a depth of about 4600 km. Most disagreements arise in the transition zone F between the outer and inner core.

The hypothesis of a rapid or discontinuous increase in the velocity of seismic P waves in the Earth's core was first suggested by Lehmann (1936). Calculations by Gutenberg and Richter (1938, 1939) and by Jeffreys (1939c, d) corroborated Miss Lehmann's hypothesis but differed in the interpretation of the velocity distribution near the inner core boundary. Jeffreys assumed that the wave velocity decreased with depth just above the inner core, and then increased discontinuously at the inner core boundary (see Fig. 1.3). Gutenberg on the other hand favoured a gradual increase in velocity in a transition zone without a preceding decrease. Neither of these models, however, can explain (using ray theory alone) the observed short-period precursors to *PKP* waves at distances less than 140°. Jeffreys (1939c) considered diffraction near a caustic and showed that appreciable amplitudes for a diffracted *PKP* wave could not exist more than about 3° from the *PKP* caustic at 143° for periods of 1 sec, or 14° for periods of 10 sec. Denson (1952) and Gutenberg (1957, 1958a, b, 1959) attributed these early arrivals to dispersion in the transition zone. The lack of definitive observational evidence, however, is a significant argument against accepting this as the mechanism that is operative in the Earth. An explanation for the early onsets in terms of magnetoelastic coupling was shown to be unlikely by Knopoff and Macdonald (1958).

More recently, Bolt (1962, 1964) proposed that two discontinuous increases

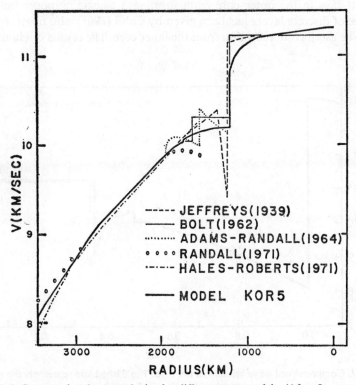

Fig. 1.6. Compressional wave velocity for different core models. (After Qamar, 1973.)

in the velocity distribution of *P* waves in the Earth's core could explain these arrivals as rays refracted through an intermediate shell. He used observations from a study of *PKP* readings in the range $110° < \Delta < 145°$ (Bolt, 1959), from four 1954 hydrogen bomb explosions and from the deep-focus Java Sea earthquake of April 16, 1957. His results indicated a discrete shell in the core with a thickness of 420 km and a mean *P* velocity of 10·31 km/sec (see Fig. 1.6). This shell surrounds the inner core with a mean radius of 1220 km and a mean *P* velocity of 11·22 km/sec. Hai (1961, 1963), in a study of arrivals from deep-focus earthquakes in the Fiji, New Hebrides and Celebes Islands, proposed a layered core with dispersion in selective layers: Bolt (1964) suggested that Hai's observations could be accounted for by his model of the core. Adams and Randall (1963, 1964), using observations of *PKP* phases from well-located earthquakes, derived travel-time curves which demanded three discontinuous jumps in the *P* velocity distribution near the inner core boundary. Their velocity solution shows two shells, each between 300 and 400 km thick, surrounding the inner core, with a small negative velocity gradient in each shell (see Fig. 1.6). The outer discontinuity is sufficiently shallow to prevent rays in the outer core from forming a caustic. Support for such a scheme of discrete layers has been given by Caloi (1961) who observed waves which he ascribed to reflection from the inner core. The results of Hannon and

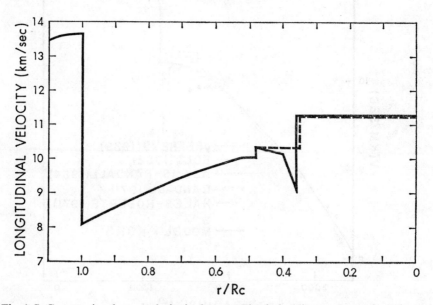

Fig. 1.7. Compressional wave velocity in the core. The dashed line represents the Bolt core model. (After Engdahl, 1968.)

Kovach (1966), who used an extended array to determine apparent velocities for *PKP* observations in the range 130°–160°, also tentatively support the Adams–Randall core model. Subiza and Bath (1964) studied *PKP* waves recorded by the Swedish seismological network and concluded that the transition zone is more complicated than previously assumed and possibly has internal layering. Ergin (1967) also analysed *PKP* travel-times and amplitudes recorded on both long and short-period seismographs and obtained a much more detailed structure of the core and upper mantle. His model includes several low velocity layers and introduces complications to *PKP* travel-times which are difficult to interpret. Jacobs (1968) has pointed out other difficulties with Ergin's model.

More recently, Engdahl (1968) has carried out a detailed study of the structure of the Earth's core, using observations of *P*, *PKP* and *pP* to determine precisely the hypocentral parameters for two well-recorded deep-focus earthquakes and a near surface event. Arrival times and amplitudes of core phases from these events were supplemented by recently published *PKP* observations. Engdahl first tried to find a solution in terms of a decreasing velocity function within the intermediate shell of the Bolt core model—his velocity distribution is shown in Fig. 1.7 (the Bolt core model is indicated by the dashed line). He obtained an alternative velocity model (Fig. 1.8) by

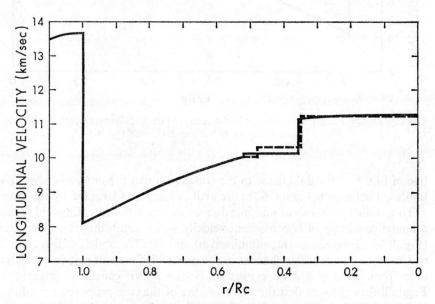

Fig. 1.8. Alternative compressional wave velocity in the core. The dashed line represents the Bolt core model. (After Engdahl, 1968.)

increasing the radius of the Bolt intermediate layer to a point just below the radius of penetration ($r_p = 1881$ km) of the point B (the *PKP* caustic). These two models represent the limits of allowable velocity solutions for a single layer transition zone corresponding to the observed data.

Engdahl (1968) also considered more complicated models with additional discontinuities in the core. The velocity function proposed by Adams and Randall (1964) consisted of three discontinuous increases between the outer and inner core—the details of this velocity structure are shown by the dashed

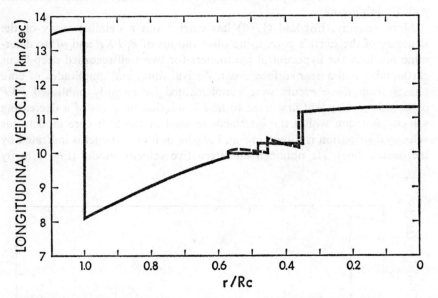

Fig. 1.9. Compressional wave velocity in the core. The dashed line represents the Adams–Randall model. (After Engdahl, 1968.)

line in Fig. 1.6. Engdahl came to the conclusion that the unusual curvatures indicated between 0·3 and 0·6 Rc are artificial and unwarranted by the data.

To test the hypothesis of multiple discontinuities in the core, Engdahl set up a model consisting of two constant velocity shells within the transition zone (Fig. 1.9). To be consistent with observational data for models of this type, it was necessary to consider changes in the core which prevent rays in the outer core from forming a *PKP* caustic at B (the *PKKP* caustic is preserved). Engdahl thus believes that the fine structure of the core proposed by Adams and Randall has not been established and demands re-evaluation.

Engdahl concluded that a working model of the core consisting of a single-layer transition zone (a modified Bolt model) can explain the observed data within reasonably close limits. This layer is between 400 and 600 km thick and has a *P* velocity distribution either constant or decreasing with depth. In order to explain early *PKP* arrivals, characteristically observed in the range 125°–140°, a velocity discontinuity must be introduced in the core between radii of 1667 and 1806 km, approximately. The radius of this discontinuity defines the top of the transition layer. The bottom of the layer or inner core boundary, also defined by a jump in the velocity, is between radii of about 1233 and 1250 km.

Haddon (1972), in a further study of precursors to *PKIKP*, concluded that they were not likely to be associated with the presence of a transition layer of the Bolt type nor with any internal discontinuities inside the outer or inner core, and suggested that they may be caused by scattering of *PKP* waves from irregularities near the MCB. In a later paper, Cleary and Haddon (1972) examined this suggestion in more detail and presented additional evidence in its favour. King *et al.* (1973) have since obtained measurements of slowness, i.e. (*dT*/*d*Δ) variations, within *PKIKP* precursor wave trains which they claim provide crucial evidence in favour of scattering near the MCB. The obvious lack of scatter in the higher-order *PnKP* arrivals observed by a number of authors suggests that lateral inhomogeneities in the lower mantle are the likely source of scattered waves rather than "bumps" on the MCB. Further support to the suggestion that precursors to *PKIKP* are the result of the scattering of *PKP* waves by random irregularities near the base of the mantle has been obtained by King *et al.* (1974) from an analysis of 12 large earthquakes in the distance range 128°–142° from the Warramunga seismic array in northern Australia. In a later, more detailed analysis, Haddon and Cleary (1974) showed that scattering of *PKP* waves by random irregularities in densities and elastic parameters of about one per cent in the region *D″* (the bottom 200 km of the mantle) is consistent with most of the evidence on precursors to *PKIKP* including travel times, slownesses, azimuthal deviations, amplitudes and amplitude variations. In addition to meeting the observational requirements, this interpretation has the further advantage of simplicity—the only discontinuity in the core being the boundary between the inner and outer core. This is also a feature of model B1 of Jordan (1972)—see Section 1.6. It must be pointed out, however, that some authors still claim that the seismic data demand a two layer transition zone in the core, although not excluding inhomogeneities at the base of the mantle which they consider a secondary effect—see, for example, Bertrand and Clowes' (1974) analysis of travel-times and travel-time gradients recorded at the Warramunga seismic array. These authors claim that the data are best fitted by three discontinuous increases in the *P* velocity, one at the outer core-inner core boundary and two at shallower depths. These discontinuities define two layers, each a few hundred km thick surrounding the inner core—the velocity increase in both of these two layers being less than 0·1 km/sec.

For most models, P velocities in the inner core are approximately constant, the velocity increase at the inner core boundary being 0·9–1·0 km/sec. The models of Buchbinder (1971) and Qamar (1973) are exceptions—they show a velocity increase of about 0·6 km/sec at the inner core boundary followed by a pronounced velocity gradient in the outermost part of the inner core. Qamar's model ($KOR5$) is also shown in Fig. 1.6—it was developed to satisfy all available observations of travel-times, $dT/d\Delta$, and amplitudes of core waves. The model is also tied to the Herrin (1968) tables for P and the Randall (1970) tables for SKS. The average velocity in the transition region is $1\frac{1}{2}$ per cent lower than in Bolt's (1962) model: at the outer boundary of the transition zone ($r = 1782$ km), model $KOR5$ has a velocity jump of only 0·013 km/sec, which is only 1/20 of that in Bolt's (1962) model. The jump at the inner core boundary ($r = 1213$ km) is 0·6 km/sec, which is about 2/3 of that in most other models.

Very similar results have been obtained by Müller (1973) from a study of long period core phases. He found that the P velocity distribution is quite smooth right down to the inner core boundary with no first-order discontinuities in the transition region. The P velocity increase at the inner core boundary is 0·6–0·7 km/sec with again a strong velocity gradient in the outer part of the inner core; the S velocity just inside the boundary of the inner core is ~ 3.5 km/sec. This is greater than the value of 2·95 km/sec obtained by Julian et al. (1972) in their paper claiming to have identified the phase $PKJKP$ and confirms the value obtained earlier by Dziewonski and Gilbert (1971) based on normal mode oscillations (see Section 1.3).

The shear velocities in the Earth's lower mantle have long been thought to increase moderately and monotonically downward to the MCB. Recently, however, some shear velocity models have been proposed in which the region just above the MCB contains a significant negative velocity gradient. Such models have been proposed by Cleary (1969), Bolt et al. (1970) and Robinson and Kovach (1972) on the basis of $dT/d\Delta$ determinations of diffracted S waves. Their estimates of the shear velocity just above the core range from between 6·81 and 6·99 km/sec, as opposed to Jeffreys' value of 7·30 km/sec.

Observations of the periods of free oscillations of the Earth provide additional data for determining the shear velocity distribution (see Section 1.3). Trade offs, however, occur between the shear velocity, density, and core radius, and a unique determination of these is difficult (Dziewonski, 1970). The effect of these trade offs can be reduced if overtones of the spheroidal and toroidal modes are incorporated in the inversion. In a recent study, Jordan (1972) combined all available free oscillation data with a large body of travel-times and found that, although a model with a negative shear velocity gradient in the lower 100 km of the mantle provided a satisfactory fit to the data, the fit was not as good as that in which the negative velocity gradient was absent.

Mitchell and Helmberger (1973) examined details of the velocity distribution near the base of the mantle that are not resolvable using free oscillation data or travel-times of reflected waves. Their data consisted of long-period S and ScS waves recorded by stations of the World-Wide Standardized Seismograph Network (WWSSN) and the Seismological Service of Canada. They compared amplitude ratios of these phases with the ratios determined from synthetic seismograms generated by the Cagniard-deHoop method for a spherically layered Earth (Gilbert and Helmberger, 1972). They found that the observed amplitude ratios of transversely polarized shear waves $ScSH/SH$ exhibited a minimum at a distance of about 68°. Synthetic seismograms computed for a Jeffreys–Bullen model and for models with negative linear velocity gradients at the base of the mantle fail to explain this feature. On the other hand various *positive* linear velocity gradients above the MCB can explain the amplitude ratio minimum as well as an apparent difference in arrival times of the transversely and radially polarized core reflections ScS. Good agreement between the observed amplitude ratios and those computed cannot, however, be achieved without assigning low Q values to the lower mantle or a small shear velocity to the outer core. High-velocity regions between 40 and 70 km thick containing increases of from 0·3 to 0·5 km/sec above the velocity of a Jeffreys velocity model yield the best explanation to the combined amplitude and differential time data. Mitchell and Helmberger also observed ScS phases that bottom at widely diverse points on the MCB. They concluded that it is likely that a high velocity zone is widely prevalent at the base of the mantle and may even be a general feature of that region. Lateral variations in this layer could easily account for the scatter in $dT/d\Delta$ measurements at large distances. Such a model is consistent with the proposed core evolution of Anderson and Hanks (1972) (see Section 2.4) in which silicate material rich in calcium, aluminum and uranium and having high velocities might remain as a residue from the early processes of core formation. The model derived by Mitchell and Helmberger contrasts sharply with diffusion models, in which it is thought that liquid core material has penetrated great distances into the lower mantle. If such a phenomenon does occur, it must be restricted to a thin layer that remains undetected by the methods they applied.

1.3. Free Oscillations of the Earth

When any part of a deformable body like the Earth is in motion, it may be regarded as either a vibrating system with an indefinite number of degrees of freedom or as a medium transmitting waves. Seismic wave theory regards the associated Earth motions as travelling disturbances affecting only a relatively small part of the Earth's volume at any given time. Oscillation theory, on the other hand, regards the motions as normal modes of oscillation, the principal ones of which affect a relatively large fraction of the Earth's volume at any instant. The wave theory approach is usually more appropriate for shorter periods (less than about 3 min.), the oscillation approach for longer periods. The relatively late application of free Earth oscillation theory is due to the fact that only recently could instruments record ground motions with periods greater than about 3 min.

Free Earth oscillations were first observed in 1952, although it was not until the great Chilean earthquake of May 22, 1960 that improved instrumentation enabled detailed measurements to be made—additional results were obtained from the great Alaska earthquake of March 28, 1964. Although the free oscillations of a uniform elastic sphere were first investigated more than a hundred years ago, calculations for realistic Earth models taking into account detailed structure, gravitational forces and rotation were not undertaken until fairly recently (see e.g. Alterman *et al.*, 1959). Such calculations would have been impossible without the aid of modern computers. The free oscillations excited by a major earthquake last for several days, but their amplitude diminishes because the Earth is not a perfectly elastic body. If spectra are computed for successive time intervals following the excitation, the damping of each mode can be determined and information on the anelasticity of the Earth's interior obtained. From an analysis of the normal modes from 84 recordings of the Alaska earthquake of March 1964, Dziewonski and Gilbert (1971, 1972) concluded that the inner core of the Earth must be solid. This was before Julian *et al.* (1972) claimed to have identified the phase *PKJKP* on seismograms.

There are two main classes of oscillations. The first class are called torsional (or toroidal) oscillations—the dilatation is everywhere zero and there is no radial component of displacement. Since the dilatation is zero, torsional oscillations cause no disturbances in density and hence no disturbances in the gravitational field. Thus instruments designed to measure small fluctuations in gravity cannot record torsional oscillations, although spheroidal oscillations may be detected. Spheroidal oscillations are coupled oscillations and involve both radial displacements as well as torsional motions. The deformation of the Earth's surface is best described by spherical harmonic functions $P_n^m (\cos \theta)_{\sin m\phi}^{\cos m\phi}$, where $P_n^m (\cos \theta)$ are the associated Legendre functions and θ, ϕ the co-latitude and longitude. Each class of vibration is characterized by the order m and degree n of the spherical harmonic involved and by the number of nodal surfaces in the radial direction. It has become general practice to specify oscillations by the symbols $_lS_n^m$ and $_lT_n^m$ for the spheroidal and toroidal types respectively, the subscript l indicating the number of nodal surfaces. The superscript m is dropped when there is no longitude dependence*. For the lowest values of n, the fundamental free Earth oscillation periods are determined predominantly by the properties of the Earth's deeper interior. For these values of n, the effect of lateral variations in the properties of the Earth on oscillation periods are negligible. For higher values of n, the periods become increasingly influenced by properties nearer the surface of the Earth where lateral variations may be more significant.

The fundamental spheroidal oscillation $_0S_0$ is an alternating compression and rarefaction of the whole Earth. There are an infinite series of overtones $_lS_0$ with spherical nodal surfaces within the Earth. The case $n=1$ is precluded —it represents a net translation of the surface and hence of the centre of gravity and would require the application of an external force. $_0S_2$ represents the "football" mode. There are two nodal parallels of latitude dividing the surface into three zones. As the sphere oscillates, it is distorted alternately into an oblate and prolate spheroid. Spheroidal modes $_0S_3$, $_0S_4$ . . . are motions with an increasing number of subdivided zonal distributions—each of these modes also has overtones with internal nodal surfaces.

For torsional oscillations, there is no solution for $n=0$. The case $_0T_1$ is precluded. It implies a variation in the rate of rotation of the whole Earth, concentric shells within the Earth executing small rigid body rotations about the polar axis. The simplest torsional mode is $_0T_2$—in this case there is a nodal line for displacement around the equator, the two hemispheres oscillating in anti-phase. Higher modes arise from subdivisions of the Earth into 3, 4 . . . zones with opposite motions. There are also overtones with internal nodal surfaces.

For Earth models which are spherically symmetric and non-rotating, it is not necessary to include the parameter m—the oscillations may be described using only the two parameters (l, n). For a rotating body it is necessary to include m. Travelling waves are set up where wave fronts are diametral planes containing the polar axis and revolving about it, either east–west or west–east depending on the sign of m. The longest free Earth oscillation periods (~ 1 hour) are small compared with the axial rotation period (1 day). The effect of axial rotation is thus small. When it can be neglected, then for any value of m, the above east–west and west–east waves combine to produce standing waves (which are in reality oscillations) in which the frequency is independent of the value of m. Thus an effect of axial rotation is to cause "split modes", i.e. a small separation of a single mode into two or more distinct modes with

* In much of the literature on the free oscillations of the Earth these symbols are written $_nS_l^m$ and $_nT_l^m$ with the subscripts (l, n) interchanged. We have however preserved the more common notation P_n^m (cos θ) for the associated Legendre functions and use l to indicate the number of nodal surfaces.

slightly different periods. Backus and Gilbert (1961) showed that axial rotation causes the nodal pattern to drift westward round the Earth's axis with angular speed equal to the frequency difference between successive members of the split mode. The Earth's ellipticity contributes further to the splitting of modes.

The records of free Earth oscillations show that the lowest frequency modes often appear as a number of closely spaced lines in the spectrum. Pekeris *et al.* (1961) suggested that this might be due to the rotation of the Earth splitting any mode having a longitude (m) dependence into a number of frequencies. They coined the term "terrestrial spectroscopy" for the study of free oscillation spectra, pointing out that the observed splitting of certain lines due to the

Table 1.2. Gravitational undertone periods of S_2^2 in hours. (After Smylie 1973.)

Radial Order	Pekeris and Accad (1972)	Coriolis free [t]	Actual Period
0*	0·89583	0·89744	0·87125
1	7·3515	7·4524	6·7598
2	11·6735	12·4716	10·5285
3	16·0187	17·5474	13·8070
4	20·2618	22·3871	16·4970
5	24·3077	26·8703	18·6584
6	—	—	20·3377

* This is the gravest elastic mode of vibration of the Earth.

[t] These periods differ from those of Pekeris and Accad (1972) presumably because of the presence of the solid inner core in the Earth model used in Smylie's study.

Earth's rotation is a mechanical analogue of the Zeeman splitting of lines of atomic spectra by a magnetic field.

While convective circulation is not possible in a core which is everywhere sub-adiabatic, very long period gravitational oscillations (undertones) are possible. Because the gravitational restoring forces of a sub-adiabatic core are weak compared to the elastic restoring forces of free oscillations, the periods of vibration exceed the gravest free oscillation periods. Although most free oscillation spectra show prominent energy peaks at periods longer than the gravest elastic oscillation (53·8 min.), only recently have attempts been made to compute the expected periods. The first investigation was by Pekeris and Accad (1972). They ignored the Coriolis force and Smylie (1973) has since shown that this can have a drastic effect on the periods. In his investigation, Smylie neglected the main magnetic field and assumed the liquid core to behave as an ideal fluid and calculated the undertones of the mode S_2^2 (azimuthal number $m=2$, degree $n=2$). Table 1.2 and Fig. 1.10 compare his

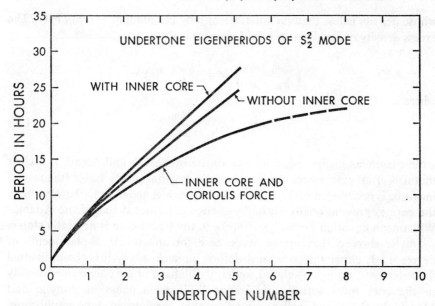

Fig. 1.10. Dispersion relations for undertone eigenperiods of S_2^2 mode. The results for the Earth model without a solid inner core are taken from Pekeris and Accad (1972). (After Smylie, 1973.)

results (for different radial orders) with those of Pekeris and Accad, and show how the periods are changed by the inclusion of the Coriolis force. Smylie has since installed a specially designed gravimeter of extremely high sensitivity to try and detect possible core undertones. Detection of such undertones (for which gravitation alone provides the restoring force) would constitute decisive evidence for stable stratification of the core and have far reaching implications for geomagnetism and the thermal history of the Earth (see Section 4.7 and 3.5).

The gravitational stability of the liquid core bears a direct relationship to the Adams–Williamson equation (1.10) which will be obtained later. If a fluid element is suddenly displaced radially by an amount δr it suffers an adiabatic expansion and its density decreases by an amount equal to its initial density multiplied by its dilation i.e.

$$\rho_0 \left(\frac{\rho_0 g_0 \delta r}{k_s} \right)$$

where (ρ_0, g_0) are the equilibrium density and gravity and k_s is the incompressibility. The displaced fluid element finds itself among neighbouring elements whose density is less than its initial density by the amount $-(d\rho_0/dr)\,\delta r$. The excess density of the displaced fluid element is thus

$$-\left(\frac{g_0\rho_0^2}{k_s}+\frac{d\rho_0}{dr}\right)\delta r=-\frac{g_0\rho_0^2}{k_s}\,\beta(r)\,\delta r$$

where

$$\beta(r)=1+\frac{k_s}{g_0\rho_0^2}\frac{d\rho_0}{dr} \tag{1.4}$$

a dimensionless factor originally introduced by Pekeris and Accad. It is clear that if $\beta(r)<0$ everywhere the liquid core is gravitationally stable (the restoring force is positive), if $\beta(r)=0$, it is neutrally stable, and if $\beta(r)>0$ everywhere, the core is gravitationally unstable. It should be noted that if the Adams–Williamson equation holds, i.e. if $\beta(r)=0$, the liquid core is neutrally stable.

Smylie showed that in the static case for spheroidal displacements of degree $n \geqslant 1$, either the core must deform in such a way that the individual fluid elements, though displaced, suffer no dilation, or the equilibrium density in the core must satisfy the Adams–Williamson equation. Smylie and Manshinha (1971) had earlier pointed out that, for fluids in static equilibrium, equipotential, isobaric and isopycnic (i.e. equal density) surfaces are parallel and that individual fluid elements may be displaced on such surfaces without resistance. Fluid elements are free to move aside on such surfaces to permit penetration of the solid inner core and mantle into the liquid outer core. Moreover such surfaces not only remain parallel but are carried radially with the fluid element.

There has been much confusion in the literature on the proper boundary conditions for the motions coupling the mantle with the liquid core, and different authors have come to different conclusions. Wunsch (1974) has shown that the essential difficulties and most of the physical phenomena can be reproduced in a much simplified model—the deformation in the static limit of an elastic half space overlying a liquid half-space. He showed that, as is often the case in fluid dynamical problems, the problem is one of singular perturbation which leads to paradoxes if not handled correctly. Wunsch showed that in the case of the coupling of the motion of an elastic mantle to a stratified fluid core, the static limit is singular in a perfect fluid and can lead to physically meaningless results. Including rotation, however, makes the static limit a regular one, since Coriolis forces can support the deformed isopycnic surfaces in a steady flow. In a later paper (1975), Wunsch extended his analyses to include the effects of viscosity and density diffusion, and showed that paradoxes and contradictions which arise by treating only perfect fluids, may be resolved by taking dissipation into account. Wunsch also considered the effect of a large ambient magnetic field—depending upon its orientation, its effect is similar to that of stratification or rotation. In particular, a toroidal field introduces a static limit singularity similar to that of density stratification.

1.4. Variation of Density and Other Physical Properties within the Earth

The density ρ will depend on the pressure p, the temperature T, and an indefinite number of parameter n_i specifying the chemical composition i.e.

$$\rho = \rho\,(p, T, n_i). \tag{1.5}$$

If m is the mass of material within a sphere of radius r, then, since the stress in the Earth's interior is essentially equivalent to a hydrostatic pressure*,

$$\frac{dp}{dr} = -g\rho \tag{1.6}$$

where

$$g = \frac{Gm}{r^2} \tag{1.7}$$

and G is the gravitational constant. Considering first a chemically homogeneous layer in which the temperature variation is adiabatic, it follows from Eqn. (1.5) that

$$\frac{d\rho}{dr} = \frac{d\rho}{dp}\frac{dp}{dr} = \frac{-g\rho^2}{k_s} \tag{1.8}$$

where k_s is the adiabatic incompressibility defined by the equation

$$\frac{1}{k_s} = \frac{1}{\rho}\left(\frac{\partial\rho}{\partial p}\right)_s \tag{1.9}$$

where s is entropy. A homogeneous region is here defined as one in which there are no significant changes of either phase or chemical composition. It follows from Eqns. (1.7), (1.8) and (1.3) that

$$\frac{d\rho}{dr} = \frac{-Gm\rho^2}{k_s r^2} = \frac{-Gm\rho}{r^2\phi}. \tag{1.10}$$

* In the deeper interior of the Earth, strains are fairly large (e.g. near the MCB, the mean of the principal strains has magnitude about 0·14). The mean ($\frac{1}{3}p_{kk}$) of the principal stresses is correspondingly large. In contrast the strength of the materials in the Earth, and therefore the greatest deviatoric stresses that can be attained, is small i.e. the ratio of any p_{ij} to $\frac{1}{3}p_{kk}$ (which corresponds to pressure) is small at depths greater ~ 50 km and diminishes steadily as the depth increases. We can thus neglect the p_{ij} compared with $\frac{1}{3}p_{kk}$ in the Earth's deep interior and treat the internal stresses of the Earth, when in its equilibrium state, as hydrostatic.

The distribution of ϕ throughout the Earth is known from the velocity–depth curves (cf Eqn. 1.3).

Since $dm/dr = 4\pi\rho r^2$, a second-order differential equation for $\rho = \rho(r)$ can be written down by differentiating equation (1.10). This equation, which was first obtained by Adams and Williamson in 1923 (Williamson and Adams, 1923), may be integrated numerically to obtain the density distribution in those regions of the Earth where chemical and non-adiabatic temperature variations may be neglected.

From Eqns. (1.6) and (1.7), it follows that

$$\frac{dp}{dr} = \frac{-Gm\rho}{r^2}. \tag{1.11}$$

Hence by numerical integration, the pressure distribution may be obtained once the density distribution has been determined. Since the density is used only to determine the pressure gradient, the pressure distribution is insensitive to small changes in the density distribution and may be determined quite accurately. The variation of g can be calculated from Eqn. (1.7). From a knowledge of the density distribution it is easy to compute values of the elastic constants. Thus Eqns. (1.2) and (1.3) give μ and k_s directly.

Any density distribution must satisfy two conditions—it must yield the correct total mass of the Earth and the moment of inertia about its rotational axis. Using these two conditions and a value ρ_1 of 3·32 g/cm^3 for ρ at the top of layer B of the mantle (assumed to be at a depth of 33 km), Bullen applied Eqn. (1.10) throughout the regions B, C and D. He then found that this led to a value of the moment of inertia I_c of the core greater than that of a uniform sphere of the same size and mass. This would entail the density to decrease with depth in the core and would be an unstable state in a fluid. Birch (1952) showed that a non-adiabatic temperature gradient only worsened the situation. A reasonable value for I_c could be obtained by increasing the initial value of ρ_1, but only if an impossibly high value (at least 3·7 g/cm^3) were chosen. Thus the assumption of chemical homogeneity must be in error; the region where this assumption is most likely to be invalid is region C where there are large changes in the slope of the velocity-depth curves. In his original Earth Model A, Bullen thus used the Adams–Williamson eqn. (1.10) in regions B and D while in region C he fitted a quadratic expression in r for $\rho = \rho(r)$.

In the outer core (region E) Eqn. (1.10) is likely to apply, and values of ρ down to a depth of about 5000 km can be obtained with some confidence. In the core one boundary condition is $m = 0$ at $r = 0$ but lack of evidence on the value of the density ρ_0 at the centre of the Earth leads to some indeterminacy in the density distribution in regions F and G. Bullen initially derived density distributions on two fairly extreme hypotheses, (i) $\rho_0 = 12\cdot3$ g/cm^3 and (ii)

$\rho_0 = 22\cdot3$ g/cm^3 (this value being taken quite arbitrarily). A model with density values midway between those of these two hypotheses has been called Model A. More recent evidence indicates that ρ_0 is probably much nearer its minimum value and that a model based on $\rho_0 = 12\cdot3$ (Model A–i) is more likely to be correct*.

These earlier determinations of the density distribution only made use of the information contained in the velocity–depth curves. The biggest source of additional information has come from analyses of the free vibrations of the Earth excited by the two major earthquakes of 1960 and 1964. None of the existing velocity–depth curves combined with Bullen's density distribution were consistent with the longer period free oscillation data. Using this additional data Landisman *et al.* (1965) investigated a number of Earth models without assuming homogeneity and an adiabatic temperature gradient (except in the outer core E). A feature of their models is constant density between depths of about 1600 and 2800 km. Bullen and Haddon (see later), were able to avoid this very implausible conclusion by treating the radius of the core r_c as a free parameter—a normal value of the density gradient in the lower mantle could be obtained by increasing r_c by about 15 km. Dorman *et al.* had earlier (1965) proposed an Earth model in which r_c was increased by 10 km. Pekeris (1966) also obtained density distributions in the Earth without assuming homogeneity and an adiabatic temperature gradient in any region of the Earth. His density distribution $\rho(r)$ was represented by 50 pivotal values $\rho(r_k)$ with linear variations in between and with discontinuities at the base of the crust and at the MCB. Pekeris varied the ρ_k by the method of steepest descent so as to minimize the sum of the squares of the residuals of all the observed periods of the free oscillations of the Earth. As might be expected the density distribution in the inner core has little effect on the spectrum as a whole.

Other workers have supplemented the travel-time data with experimental empirical data. Birch (1964) used an approximate linear relationship between density and the velocity of compressional waves which he had observed for silicates and oxides of about the same iron content viz.

$$\rho = a(\bar{m}) + bV_P \tag{1.12}$$

where $\bar{m}$ is the mean atomic weight. Birch only used this equation in the upper mantle (where there are high thermal gradients) and the transition zone (where there are phase changes). In the rest of the Earth he used the Adams–Williamson equation. Wang (1970) suggested that a better estimate of the density in the upper mantle might be obtained by using an empirical relationship between the bulk sound velocity $c = (V_P^2 - \frac{4}{3}V_S^2)^{1/2}$ and density ρ rather than Birch's relation (1.12) between V_P and ρ.

* A good account of this early work can be found in Bullen (1963).

B

Chung (1974) has obtained new ultrasonic data on some crustal structures important in mantle minerology, using a powder-matrix method (Simmons and Chung, 1968) in which the elastic parameters of the unknown material are inferred from the measured elastic parameters of a composite made of the powdered material imbedded in a vacuum hot-pressed matrix of AgCl. Chung showed that a power law describes the velocity–density (V_P, ρ) relationship for high pressure polymorphs better than Birch's linear law. In particular he found that V_P varied quite linearly with density over very wide density (and thus wide pressure) ranges for materials of lower mean atomic weight, the linearity even being preserved through solid–solid phase transformations. For materials of higher mean atomic weight on the other hand evidence for linearity is weak. In a later paper Shankland and Chung (1974) obtained a power law equation for the dependence of V_P upon $\bar{m}$ and ρ. Birch's linear law and Anderson's (1967) seismic equation of state may be deduced from it as special cases.

Clark and Ringwood (1964) and Wang (1972) have estimated densities in the Earth using petrological models for the upper mantle. Mizutani and Abe (1972) obtained an Earth model consistent with all the data using a trial and error method with the help of an equation of state of some rock-forming minerals.

It must be emphasized that the overall density distribution within the Earth is not changed drastically by taking into account the additional observational data on its free vibrations and the revised estimate of its moment of inertia about its axis of rotation as determined from analyses of the orbits of artificial satellites (Cook, 1963). Inside the mantle the largest difference in ρ between models A–i and HB$_1$ (to be discussed later) is only 0·15 g/cm^3, while inside the core the values of ρ in model HB$_1$ exceed those in model A–i at all depths by amounts between 0·2 and 0·3 g/cm^3.

Buchbinder (1968) carried out a detailed study of PcP phases from eight explosions and three earthquakes. He reported cases of reversed polarity of PcP phases relative to P for epicentral distances $< 32°$ at which distance the amplitudes of PcP pass through a minimum. Buchbinder suggested that these results reflect properties at the MCB which cannot be satisfied by any of the "conventional" Earth models. He found that the velocity at the top of the core is some 7–8 per cent lower than values usually quoted and that there is no density discontinuity across the MCB. Buchbinder interpreted his results to indicate that the bottom of the mantle is inhomogeneous—the inhomogeneity being caused by an increase in iron (or other heavy metal) content with depth which increases the mean atomic weight and density with but little change in P velocity. The drop in P velocity across the MCB may then be explained by a discontinuous increase in mean atomic weight with but little change in density.

Various authors have objected to Buchbinder's interpretation. Berzon *et al.* (1972) noted that *PcP* phases are observed near 30° as frequently as at other epicentral distances, thus ruling out a zero in amplitude. Kogan (1972) demonstrated, by correlation between *P* and *PcP* phases, that *PcP* phases do not invert their polarity. Chowdhury and Frasier (1973) carried out an analysis of short period *P* and *PcP* phases from earthquakes in the range $26° < \Delta < 40°$ from the Large-Aperture Seismic Array (LASA) in Montana. No reversals of *PcP* polarity for $\Delta < 32°$ and no minimum in amplitudes at 32° were observed—again ruling out Buchbinder's model and favouring the more conventional models with a density ratio at the MCB of about 1·7. Buchbinder himself has now abandoned his original interpretation.

A new approach to the inverse problem is the Monte Carlo method described in its geophysical context by Keilis-Borok and Yanovskaya (1967). It was used by Press (1968a, b) to obtain a number of Earth models using as data 97 eigenperiods, travel-times of *P* and *S* waves and the mass and moment of inertia of the Earth. The Monte Carlo method uses random selection to generate large numbers of models in a computer, subjecting each model to a test against geophysical data. Only those models whose properties fit the data within prescribed limits are retained. This procedure has the advantage of finding models without bias from preconceived or over simplified ideas of Earth structure. Monte Carlo methods also offer the advantage of exploring the range of possible solutions and indicate the degree of uniqueness obtainable with currently available geophysical data. Press was able later (1970a, b) to speed up considerably his Monte Carlo procedures. Using new and more accurate data he was able to find a large number of successful models—Fig. 1.11 and Fig. 1.12 show 27 successful density distributions in the mantle and core respectively. Press also confirmed that the Earth's core is inhomogeneous and that the density at the top of the core is constrained to the narrow range 9·9–10·2 g/cm³. Wiggins (1969) also used Monte Carlo techniques to investigate the nature of the non-uniqueness inherent in the interpretation of body wave data.

Haddon and Bullen (1969) have constructed a series of Earth models (HB) using free oscillation data consisting of the observed periods of fundamental spheroidal and toroidal oscillations for $0 \leqslant n \leqslant 48$ and $2 \leqslant n \leqslant 44$ respectively and certain spheroidal overtones. Data from the records of both the Chilean (May, 1960) and Alaskan (March, 1964) earthquakes were used. Their procedure was to start from models derived independently of the oscillation data and to produce a sequence of models showing improved agreement with all the available data. In passing from one model to the next a guiding principle was to introduce and vary one or more of the parameters in the model description at any stage in order to satisfy the oscillation data. They thus tried to establish models described in terms of the minimum number of parameters

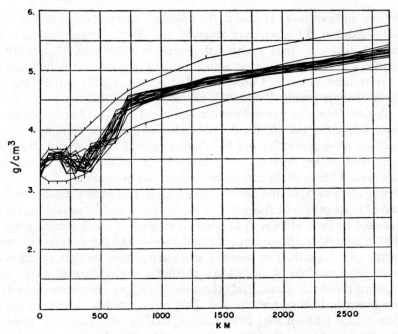

Fig. 1.11. Successful density distributions in the mantle. (After Press, 1970a.)

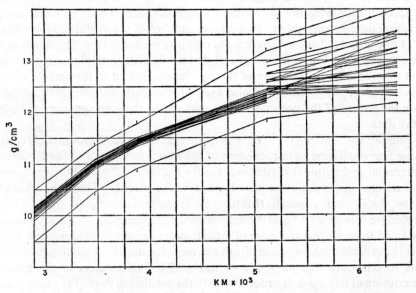

Fig. 1.12. Successful density distributions in the core. (After Press. 1970a.)

demanded by the data. A major difference in principle between their method and the Monte Carlo procedure of Press is the comparatively large number of parameters that Press permits to be randomly varied. Haddon and Bullen pointed out that the predominance of complex models found by Press is inherent in his method, since a simple random walk would automatically have a low probability. Haddon and Bullen also stressed that the "average" Earth to which average periods of free oscillation modes relate is not necessarily the same as an Earth model to which the currently available average seismic body wave travel-times apply since earthquake epicentres and recording stations are not randomly distributed over the Earth's surface.

In 1970, Bullen and Haddon used additional evidence to derive an improved Earth model (HB2). Model HB2 incorporates the newer P travel-time data of Herrin (1968) and takes into account the abnormalities in the body wave observations in the lower 200 km of the mantle and the detailed structure of the lower core (Bullen, 1965). In their earlier model HB1 a simplified core structure was assumed (the whole core being considered fluid) since free Earth oscillation data are incapable at present of resolving fine detail in the lower core. Derr (1969) has also developed a series of Earth models using free oscillation data.

Anderssen and Seneta (1971, 1972) have examined in detail the Monte Carlo method of inversion of geophysical data: in particular they developed a statistical procedure for estimating the reliability of non-uniqueness bounds defined by a family of randomly generated models. Anderssen et al. (1972) have applied these techniques to the problem of obtaining the density distribution within the Earth, and in a later paper (Worthington et al., 1972) attempted to resolve the major discrepancies between the density models of Bullen and Haddon and those of Press. They showed that these differences cannot be due to the different techniques employed to derive the models, and concluded that they are predominantly due to differences in assumptions about the permissible range of values of the shear velocity in the upper mantle.

An upper bound to the density increase at the boundary of the inner core has been obtained by Bolt and Qamar (1970) from measured amplitudes of PKiKP and PcP phases recorded at LASA in Montana. The PKiKP waves were reflected at steep angles from the inner core boundary (epicentral distance $\Delta < 40°$) and demonstrate the sharpness of that boundary (Engdahl et al., 1970). Bolt and Qamar's analysis indicated a minimum value of 0·875 for the ratio of the densities at the inner core boundary. This would give a maximum density jump $\Delta\rho_{ic}$ at the inner core boundary of 1·8 g/cm³ if the density in the outer core at the boundary is 12·35 g/cm³ (see also Bolt, 1972). This result has been obtained on seismological evidence alone and does not depend on any assumptions concerning the chemical composition of the core.

It is also in agreement with recent core models which indicate that the density in the inner core is much lower than was at one time supposed.

Dziewonski (1971) has stressed that only observations of overtones will allow the density distribution to be determined with sufficient detail to be able to make meaningful estimates of the structure and composition of the Earth's deep interior. The diversity of models which satisfy travel-times and fundamental mode data indicates the insufficiency of the constraints provided by the more limited set of data. Dziewonski was able to identify overtones with periods greater than 250–300 sec on records of the March 1964 Alaska earthquake from stations belonging to the World Wide Standard Seismograph Network. Using this additional information, he constructed a number of Earth models. His final models indicate a change in composition in the bottom 500 km of the mantle. His density distributions show systematic differences from the HB_1 model and Press' models in the lower mantle, although all models lie within Press' band of solutions—except that lower shear velocities are found in the depth range 480–650 km. Although agreement with these other models is good for the fundamental modes and the first two spheroidal overtones, it is poor for the third and fourth spheroidal and torsional overtones. A solid inner core is also demanded by Dziewonski's data—if the shear velocity in the inner core is 3·5 km/sec, $\Delta\rho_{ic} = 0·81$ g/cm³.

1.5. The Inverse Problem in Geophysics

The mathematical formulation of the inverse problem characterizes possible variations of physical parameters as entities in an abstract function space, each entity representing an Earth model. In particular a spherically symmetric, non-rotating, linearly elastic, isotropic Earth can be described by specifying the compressional velocity, the shear velocity, and the density as functions of the radius. An observation is the value of a functional defined in this space of Earth models—examples are the Earth's mass, moment of inertia, the measured travel-times of seismic waves, and the observed periods of free oscillation. Since the distributions of physical parameters are continuous in some interval and the number of data obtainable is necessarily finite, the inverse problem generally has no unique solution. Furthermore, the observations used as data are invariably contaminated by errors; only *estimates* of the values of data functionals for the Earth are available.

A general discussion of the inverse problem in geophysics has been given in a series of papers by Backus and Gilbert. They investigated (1967) the extent to which a finite set of data functionals (called by them gross Earth data) can be used to determine the Earth's internal structure. In a later paper (1968) they showed how to determine the shortest length scale which the given data can resolve at any particular depth. The principal result of their work is that it is possible to draw rigorous conclusions about the internal structure of the Earth from a finite set of gross Earth data. Infinite resolution can never be achieved yet nevertheless rigorous answers can be given to a number of qualitative questions such as whether there are low velocity zones or density inversions in the mantle. In these first two papers observational errors were neglected. In a later paper, Backus and Gilbert (1970) considered the effect of such errors and investigated the inversion of a finite set G of inaccurate gross Earth data—they showed that from some sets G it is possible to determine the structure of the Earth (except for fine-scale detail) within certain limits of error. They also showed how to determine whether a given set G will permit the construction of such localized averages of Earth structure, and how to find the shortest length scale over which G gives a local average structure at a given depth if the variance of the error is to be less than a given amount.

It is impossible to summarize briefly the theory developed by Backus and Gilbert—a very detailed account of their work has been given by Backus himself (1971). Parker (1970) has applied their method to the inverse problem of the determination of the electrical conductivity in the mantle and prefaces his paper with an outline of the general ideas and philosophy behind this extremely important work of Backus and Gilbert.

A workshop on the mathematics of profile inversion was held at NASA Ames Research Centre, California in July 1971 in which problems from a number of different disciplines were discussed. Parker (1972a) gave a brief account of the Backus-Gilbert technique and the following short summary is taken from his paper.

Consider the problem of determining a property in the Earth as a function of radius, assuming no angular dependence and consider first a linear inverse problem i.e. the observations depend in a linear way on the property. Suppose the data consist of the N real numbers γ_j ($j=1, 2, \ldots N$) presumed to be exact i.e. there is no experimental error. Then we can write

$$\gamma_j = \int_0^a m(r)\, G_j(r)\, dr, \qquad j=1, 2, \ldots N \qquad (1.13)$$

where a is the mean radius of the Earth, $m(r)$ the required property and $G_j(r)$ is called a data kernel (one for each observation). Consider, for example, $m(r)$ to be the density $\rho(r)$ and γ_j the radial component of gravity at radius r_j ($r_j \geqslant a$). Then

$$\gamma_j = \int_0^a \frac{4\pi G r^2}{r_j^2}\, \rho(r)\, dr \qquad (1.14)$$

and hence

$$G_j = \frac{4\pi G r^2}{r_j^2}. \tag{1.15}$$

For more general radial properties $m(r)$, the only thing known about m is the set of measurements γ_j and the problem is how to localize this information to points within the Earth. Backus and Gilbert consider a linear combination of γ_j given by

$$L = \sum_{j=1}^{N} a_j \gamma_j. \tag{1.16}$$

From the assumption of linearity, we have

$$L = \int_0^a \left[\sum_{j=1}^{N} a_j G_j(r) \right] m(r) \, dr. \tag{1.17}$$

If the constants a_j could be chosen so that the function in square brackets was a Dirac delta function centred on r_0, then L would be simply $m(r_0)$, the property required at the point r_0. In general it is not possible to do this and the problem becomes one of choosing the coefficients a_j to obtain a function that is concentrated strongly at r_0. Backus and Gilbert do this by defining a numerical measure of difference from a delta function and then minimizing this measure by varying the coefficients a_j. In particular they investigated the two measures

$$D_1 = \int_0^a [\delta(r - r_0) - F(r)]^2 \, dr \tag{1.18}$$

and

$$D_2 = \int_0^a F(r)^2 \, (r - r_0)^2 \, dr \tag{1.19}$$

with

$$\int_0^a F(r) \, dr = 1.$$

It is clear that if D_1 or D_2 can be made small, the function $F(r)$ will have a large peak at r_0 and unit area under it. Other measures are possible but the above two are useful since they yield simple equations for the coefficients a_j. Having obtained for every radius r_0 a set of a_j, an estimate of the property $m(r_0)$ is given by

$$\langle m(r_0) \rangle = \sum_{j=1}^{N} a_j \gamma_j. \tag{1.20}$$

If there is detailed structure with wavelengths less than the width of the chosen "delta function" at a particular depth, the measurements will not reveal it, i.e. the finite data set (although perfectly accurate) give only a smoothed version of the actual structure. In the simple example of gravity discussed above, it is clear that measurements at different r_j yield no new information about $\rho(r)$ since any linear combination of the G_j's is still proportional to r^2. Thus any radially structured model with the correct mass will satisfy the data.

In practice each of the measurements will be subject to some error so that there are two measures of imprecision (an error of the estimate and its resolution), both of which one would like to make as small as possible. It is not possible, however, to choose the a_j to minimize both at once. The error estimate of a property can be improved but only at the expense of the resolving power and vice versa. Thus there exists a tradeoff between error and resolution at every radius, and some compromise must be made in choosing the best model.

Parker (1972b) has also considered the inverse problem with grossly inadequate data. Such is the case, not only when the number of observations is small, but also if the inverse problem is intrinsically non-unique as is the case of attempting to determine the density inside a body from gravity observations outside the body. Parker showed that although inadequate data cannot yield detailed structure, nevertheless they can be used to rule out certain classes of structures and provide bounds on acceptable models. The interpretation of inaccurate, insufficient and inconsistent data has also been investigated by Jackson (1972). The question of uniqueness has been further discussed by Dziewonski (1970) and the general linear inverse problem reviewed by Wiggins (1972).

Finally it must be emphasized that most inverse problems in geophysics are non-linear and there is usually no explicit relationship between γ_j and $m(r)$. In such cases Backus and Gilbert linearize the problem. Basically, their method employs an iterative perturbation algorithm that approximates the difference between the sought representation of the Earth and some initial model as a particular solution to the finite system of linear, inhomogeneous, integral equations relating changes in the data. The data functionals are computed for the starting model and subtracted from the observed data; the system of perturbation equations is solved, and the calculated perturbation added to the starting model. This process is iterated until the data are satisfied. Because the inverse problem is non-linear and has no unique solution, interpretation of any numerical results needs special care. A common mistake is to infer that because a certain model satisfies the data, some feature of that model actually exists in the Earth, when in reality the data do not require this feature. Moreover there is no guarantee that other starting models do not exist which are outside the scope of a linear description. Recent work by Wiggins *et al.* (1973) has indicated that body wave behaviour is too

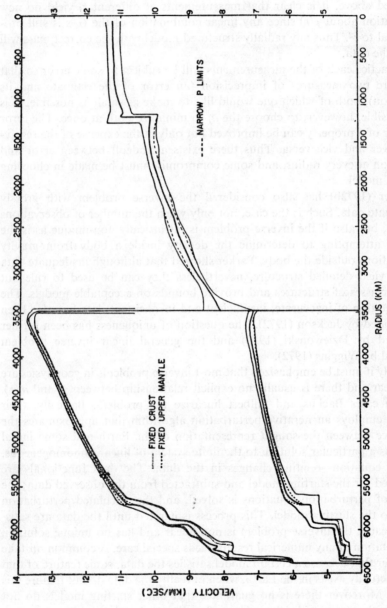

Fig. 1.13. Envelopes of all possible models of Earth structure from a given set of body wave observations. The envelope for the core was found under the assumption of a fixed core radius of 3481 km. The dark middle line is the standard model. (After Wiggins et al., 1973.)

strongly non-linear for linearized schemes to be effective in predicting uncertainties.

McMechan and Wiggins (1972) have developed a direct method for inverting seismic travel-time (T, Δ) data to determine velocity–depth profiles. They showed how a set of data points in the (T, Δ) plane with estimates of the uncertainties may be converted into an equivalent envelope in the velocity–depth plane. They refer to the problem of finding such envelopes of all possible models consistent with the observations as "extremal inversion". They believe that the extremal inversion technique makes the usual Monte Carlo inversion method obsolete. In a later paper, Wiggins *et al.* (1973) examined the range of Earth structures that can be obtained from a set of body-wave observations. Figure 1.13 shows the envelope of all possible models consistent with the set of data they used. In particular they found that there is an uncertainty of ± 40 km in the radius of the inner core boundary and ± 18 km at the MCB. The velocity uncertainty is about ± 0.08 km/sec for P and S waves in the lower mantle and about ± 0.10 km/sec in the core. An interesting result of their investigation is that quite crude observations of *SKKS–SKS* travel-times restrict the range of possible models far more than do the most precise estimates of *PnKP* travel-times.

The basic concept in the theory of Backus and Gilbert is that, although the exact solution cannot be computed because the information provided by the data is insufficient, it is possible to estimate accurately *linear averages* of the desired model. The aim of their theory is the construction of an optimal inverse filter from the constraints imposed by the observations, through which the correct solution may be viewed. As already mentioned, there is a tradeoff between the ability to resolve detail and the accuracy with which this detail can be estimated.

Jordan (1972) has used a variation of the Backus–Gilbert theory incorporating the stochastic inverse theory of Franklin (1970). A particular, unique solution to the linear system is obtained by minimizing a specified quadratic measure of error. This quadratic form is the sum of two terms, a measure of the resolution of the estimate and a measure of its accuracy, parameterized to yield a Backus–Gilbert-type tradeoff curve. Jordan showed that the generalized inverse of Penrose (1955) and Moore (1920) and the stochastic inverse of Franklin (1970) lie on this tradeoff curve, the stochastic inverse being, in one sense, an optimal point. Any particular solution computed by selecting a point on the tradeoff curve is shown to be an estimate of the correct solution convolved with a projection-like smoothing operator.

1.6. Models of the Earth's Deep Interior

Our knowledge of the region D″, the lowest 200 km of the mantle, has been reviewed by Cleary (1974). Two quite different types of models have been proposed:

(a) those in which the P and S velocities vary smoothly down to the MCB without any extreme change in gradient;
(b) those in which the velocity gradients decrease fairly abruptly at a height of ~ 100 km above the MCB and maintain a value close to the critical gradient down to the boundary.

Type (a) is represented by model $UTD124A'$ of Dziewonski and Gilbert (1972) and model B1 of Jordan and Anderson (1974). Both models are in good agreement with most travel-time and free oscillation data. Their validity rests on the supposition, supported in part by theoretical studies, that data which suggest the presence of a low velocity zone in D'' result from distortion of seismic waves by the MCB.

On the other hand, slowness and amplitude data from short period P waves indicate a fairly rapid decrease in velocity gradient at a depth corresponding to an epicentral distance of about 92°, and it is very unlikely that these data can be interpreted as interface phenomena. The measured P and S times at distances beyond about 96° also indicate reduced velocities in D''.

Type (b) is represented by model B2 of Jordan (1972), Bolt's (1972) model, and a new model designated ANU2 (Cleary, 1974). All models have high density gradients indicative of inhomogeneity in the region. Model B2 fits the oscillation data reasonably well, but has an unjustifiably low S velocity at the MCB.

Much additional information has been obtained from free oscillation data. However, problems of interpretation have arisen because of high correlations between various parameters—in particular between shear velocity and density in the lower mantle and the radius of the core (Dziewonski, 1970). Because of this, different investigators have been able to satisfy the free oscillation data with models differing radically in lower mantle structure (see Section 5.3 and Table 5.1 which gives density gradients in the lowest part of the mantle).

There is no question about the sharpness of the MCB, although there may be minor lateral variations in some of the physical properties there. Dziewonski and Haddon (1974) have recently reviewed all data on the radius r_c of the core. The first close estimate was made by Gutenberg (1913) who obtained a value of 3471 km from a determination of the distance at which P waves become diffracted. Much later, Jeffreys (1939a) proposed a method using the travel-times of core reflections and obtained a value of 3473 (± 4.2) km which he revised later (1939b) to 3473·1 (± 2.5) km. His method (with some modifications) is still being used and it is only in recent years that additional data have forced revision of his estimate.

Using travel-times of *PCP* phases from nuclear explosions, Taggart and Engdahl (1968) obtained a value of 3477 ($\pm 2\cdot0$) km for r_c. There is now convincing evidence that there are substantial lateral variations (of the order of at least 1 per cent) in V_P in the bottom few hundred km of the mantle (see, e.g. Julian and Sengupta, 1973). Earlier, Ergin (1967) and Bolt (1970) found a negative velocity gradient in the mantle just above the MCB. Bolt later (1972) inferred a 2 per cent decrease in V_P through the lowest 150 km of the mantle leading to a value for r_c of 3475 ($\pm 2\cdot0$) km. Hales and Roberts (1970a) also suggested a decrease in V_S (of about 3 per cent) in a transition shell above the MCB. In a later paper (1970b) they obtained values of r_c of 3490 ($\pm 4\cdot7$) km and 3486 ($\pm 4\cdot6$) km for two possible mantle models, basing their calculations on the differences in travel-times between *ScS* and *S* in the distance range $48° < \Delta < 70°$. More recently Engdahl and Johnson (1972) have analysed the differential travel-times *PcP–P* from three nuclear explosions in the Aleutian Islands and concluded that r_c should be increased from 5 to 15 km over the value of 3477 km obtained by Taggart and Engdahl (1968). Their preferred estimation is 3482·2 ($\pm 2\cdot9$) km.

Data of a different kind, the periods of free oscillations of the Earth, became available for the first time from an analysis of a number of records of the great Chilean earthquake of May 22, 1960 (see Section 1.3). Information obtained through observations of free oscillations of the Earth is important, not only because of the sensitivity of their eigenperiods to the density distribution, but also because of their property of averaging lateral inhomogeneities in Earth structure. Thus the normal mode method provides absolute information on the properties of a spherically symmetric average Earth; no station or source corrections are necessary. This property of normal modes makes them ideally suited to become a reference data set for derivation of a Standard Earth Model.

The free oscillation data demanded an increase in r_c over earlier estimates. Dorman *et al.* (1965) proposed a model in which r_c was increased by 10 km and a "soft" layer introduced at the base of the mantle. Haddon and Bullen (1969) later showed that the travel-time and free oscillation data could be reconciled with an Adams–Williamson density gradient (see Section 1.4) in the lower mantle by increasing r_c by 15–20 km over Jeffreys' (1939b) value of 3473 km. Press used a Monte Carlo method to generate a large number of Earth models (see Section 1.4). The successful models in his most recent study (1970b) did not show a strong preference for any particular value of the core radius within the bounds from 3463 to 3483 km.

Dziewonski (1970) later showed that only additional overtone data could narrow the range of permissible solutions based on fundamental mode data alone. Using the eigenperiods of 70 spheroidal and toroidal overtones from the spectra of 84 recordings of the Alaska earthquake of 28 March, 1964 (identified and measured by Dziewonski and Gilbert, 1972), he was able to narrow the range of r_c to 3486–3491 km. Many more overtone data have now been obtained (Dziewonski and Gilbert, 1973b—full report in preparation) and have been used by a number of authors in constructing Earth models. Two are particularly relevant to the establishment of a Standard Earth Model—that of Jordan and Anderson (1974) and that of Gilbert *et al.* (1973) —in both cases all available information with regard to the Earth's deep interior was used. Jordan and Anderson obtained a value for r_c of 3485 km and Gilbert *et al.* values of 3482·6 and 3484·9 km for two different models. The best available estimate at present for r_c is 3485 ($\pm$3) km. Although the average depth to the MCB is thus now fairly well determined, the detailed velocity structure and the physical and chemical properties of the lowest 200 km of the mantle are still very imperfectly known.

A radius of 1250 km postulated for the inner core in the Jeffreys–Bullen model now appears to be somewhat too great. Bolt's (1962) model used a radius of 1216 km and later work by Bolt (1964) supported by wide-angle reflections of *PKiKP* (Bolt *et al.*, 1968) gave values near 1220 km. The models of Buchbinder (1971) and Qamar (1973) incorporate inner core radii of 1226 and 1213 km respectively. Recent work by Engdahl *et al.* (in preparation), using *PKiKP* times and models of the earth inferred from gross Earth data, suggests an inner core radius of 1227·4 $\pm$ 0·6 km. A value in the range 1215–1234 km now seems preferable to the earlier value of 1250 km.

For the construction of his starting models, Jordan (1972) used, in the upper mantle, the linear relationship between compressional *P*-wave velocity and density proposed by Birch (1961) (see Eqn. (1.12)). The invariance of this

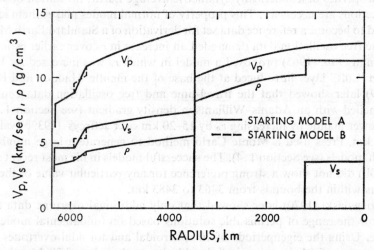

Fig. 1.14. Two starting models for the distribution of compressional velocity, shear velocity and density in the Earth. (After Jordan, 1972.)

relationship to temperature and pressure variations has been discussed by Anderson *et al.* (1970). Densities in the lower mantle and core were derived using the Adams–Williamson equation. By fixing the density at the base of the crust and fitting the mass and moment of inertia, the density profile is uniquely determined once the velocities are chosen. This is the same procedure used by Birch (1964) to construct his model II. Details of two starting models (Model A and Model B) used by Jordan are given in Fig. 1.14. He obtained three estimates of the radial distribution of compressional velocity, shear velocity and density, using an extensive set of eigenperiod and differential travel-time data. One model (B1) fits 127 of the 177 eigenperiods of the Dziewonski–Gilbert (1972) set within their 95 per cent confidence intervals, as well as extensive sets of additional data. Details of this model are given in Table 1.3 and Fig. 1.15. The radius of the core, fixed by *PCP–P* times, is 3485 km, and the radius of the inner core–outer core boundary is 1215 km. There are no other first-order discontinuities in the core; this is in agreement with the model of the core proposed by Haddon (1972) based on his interpretation of *PKP* precursors in terms of scattering near the MCB. A further account of this work has been given by Jordan and Anderson (1974).

Dziewonski and Gilbert (1973a) have extended their earlier (1972) study on the identification of the normal modes from the spectra of 84 recordings of the great Alaska earthquake of March, 1964. They identified 86 new spheroidal overtones with overtone numbers ranging from 3 to 25. In over 60 per cent of the cases, the identity of the modes was confirmed by data from the Columbian earthquake of July 31, 1970. The remaining modes were too weakly

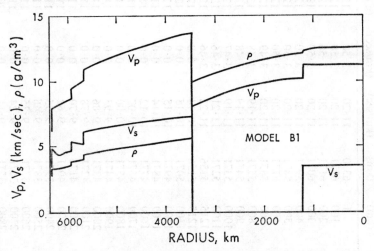

Fig. 1.15. Compressional velocity, shear velocity and density in the Earth. (After Jordan, 1973.)

Table 1.3. Details of Model B1. (After Jordan 1973.)

I	Radius (km)	Depth (km)	V_P (km/sec)	V_S (km/sec)	ρs (g/cm³)	ϕ (km²/sec²)	μ (kb)	k (kb)	λ (kb)	σ	p (kb)	g (cm/sec²)
1	1	6370	11·20	3·50	12·58	109·11	13721	1540	12694	0·4459	3609	0
2	100	6271	11·20	3·50	12·57	109·11	13716	1541	12689	0·4459	3606	52
3	200	6171	11·20	3·50	12·56	109·12	13700	1539	12674	0·4458	3598	78
4	300	6071	11·20	3·50	12·53	109·14	13671	1535	12648	0·4459	3586	110
5	400	5971	11·20	3·50	12·52	109·16	13665	1532	12643	0·4459	3570	144
6	500	5871	11·20	3·50	12·51	109·18	13663	1531	12642	0·4460	3550	178
7	600	5771	11·20	3·50	12·51	109·20	13658	1528	12639	0·4461	3525	212
8	700	5671	11·20	3·49	12·50	109·19	13649	1523	12633	0·4462	3496	247
9	800	5571	11·19	3·48	12·50	109·12	13644	1517	12632	0·4464	3463	281
10	900	5471	11·19	3·48	12·49	109·09	13630	1510	12623	0·4466	3426	316
11	1000	5371	11·19	3·47	12·46	109·22	13609	1499	12610	0·4469	3384	350
12	1100	5271	11·21	3·46	12·39	109·57	13571	1485	12581	0·4472	3339	385
13	1215	5156	11·22	3·46	12·28	109·87	13492	1467	12513	0·4475	3281	423
14	1215	5156	10·14	0·0	12·11	102·91	12460	0	12460	0·5000	3281	423
15	1300	5071	10·15	0·0	12·08	102·99	12444	0	12444	0·5000	3236	450
16	1400	4971	10·15	0·0	12·04	103·06	12411	0	12411	0·5000	3180	482
17	1500	4871	10·14	0·0	11·99	102·85	12334	0	12334	0·5000	3120	514
18	1600	4771	10·12	0·0	11·93	102·39	12219	0	12219	0·5000	3056	546
19	1700	4671	10·07	0·0	11·87	101·47	12042	0	12042	0·5000	2990	578
20	1800	4571	10·00	0·0	11·80	100·08	11805	0	11805	0·5000	2919	609
21	1900	4471	9·93	0·0	11·72	98·65	11561	0	11561	0·5000	2846	640
22	2000	4371	9·86	0·0	11·64	97·14	11307	0	11307	0·5000	2769	671
23	2100	4271	9·78	0·0	11·56	95·59	11048	0	11048	0·5000	2690	701
24	2200	4171	9·70	0·0	11·47	94·00	10785	0	10785	0·5000	2607	731
25	2300	4071	9·62	0·0	11·39	92·57	10542	0	10542	0·5000	2522	760
26	2400	3971	9·55	0·0	11·30	91·23	10309	0	10309	0·5000	2434	790
27	2500	3871	9·46	0·0	11·21	89·51	10032	0	10032	0·5000	2343	818
28	2600	3771	9·35	0·0	11·11	87·49	9718	0	9718	0·5000	2250	846
29	2700	3671	9·24	0·0	11·00	85·35	9388	0	9388	0·5000	2155	874
30	2800	3571	9·11	0·0	10·88	82·92	9023	0	9023	0·5000	2058	901
31	2900	3471	8·96	0·0	10·76	80·21	8628	0	8628	0·5000	1959	928
32	3000	3371	8·79	0·0	10·62	77·30	8209	0	8209	0·5000	1858	954
33	3100	3271	8·63	0·0	10·48	74·41	7797	0	7797	0·5000	1756	979

34	3200	3171	8·46	0·0	10·33	71·62	7400	0	7400	0·5000	1653	1003
35	3300	3071	8·31	0·0	10·19	68·99	7026	0	7026	0·5000	1549	1026
36	3400	2971	8·16	0·0	10·04	66·52	6676	0	6676	0·5000	1444	1049
37	3485	2886	8·02	0·0	9·90	64·36	6373	0	6373	0·5000	1354	1068
38	3510	2886	13·67	7·27	5·58	116·38	6489	2948	4523	0·3027	1354	1068
39	3550	2861	13·67	7·27	5·56	116·35	6466	2934	4510	0·3029	1340	1064
40	3625	2921	13·66	7·26	5·54	116·32	6443	2916	4498	0·3033	1316	1059
41	3700	2746	13·63	7·22	5·50	116·09	6385	2871	4471	0·3045	1272	1049
42	3775	2671	13·57	7·19	5·46	115·28	6294	2822	4412	0·3049	1229	1041
43	3850	2596	13·49	7·16	5·42	113·67	6160	2775	4309	0·3041	1187	1034
44	3925	2521	13·40	7·13	5·38	111·84	6014	2730	4194	0·3029	1145	1027
45	4000	2446	13·31	7·09	5·34	109·99	5870	2686	4079	0·3015	1104	1021
46	4075	2371	13·22	7·06	5·30	108·16	5727	2642	3966	0·3001	1063	1016
47	4150	2296	13·13	7·03	5·26	106·36	5597	2601	3863	0·2988	1023	1011
48	4225	2221	13·03	6·99	5·22	104·68	5469	2555	3765	0·2979	984	1008
49	4300	2146	12·95	6·96	5·19	103·12	5347	2509	3674	0·2971	944	1004
50	4375	2071	12·86	6·92	5·15	101·55	5227	2465	3583	0·2962	905	1001
51	4450	1996	12·77	6·88	5·11	99·84	5102	2422	3487	0·2951	867	999
52	4525	1921	12·68	6·85	5·07	98·13	4978	2380	3391	0·2938	829	997
53	4600	1846	12·59	6·81	5·04	96·63	4866	2336	3309	0·2931	791	996
54	4675	1771	12·50	6·77	5·00	95·13	4755	2293	3226	0·2923	754	994
55	4750	1696	12·41	6·74	4·96	93·49	4638	2253	3136	0·2909	716	994
56	4825	1621	12·33	6·71	4·92	92·10	4535	2214	3058	0·2901	680	993
57	4900	1546	12·25	6·67	4·89	90·62	4427	2174	2977	0·2890	643	993
58	4975	1471	12·16	6·64	4·81	88·99	4313	2136	2888	0·2874	607	993
59	5050	1396	12·06	6·59	4·77	87·59	4211	2089	2819	0·2872	571	993
60	5125	1321	11·97	6·55	4·72	86·07	4103	2044	2740	0·2864	535	993
61	5200	1246	11·86	6·50	4·68	84·34	3984	1998	2651	0·2851	500	994
62	5275	1171	11·75	6·47	4·64	82·36	3854	1956	2549	0·2829	465	994
63	5350	1096	11·64	6·44	4·59	80·15	3715	1920	2435	0·2796	430	995
64	5425	1021	11·52	6·39	4·55	78·30	3594	1871	2346	0·2781	395	996
65	5500	946	11·39	6·33	4·50	76·31	3469	1824	2253	0·2763	361	997
66	5550	871	11·26	6·28	4·47	74·19	3340	1778	2154	0·2739	327	998
67	5600	821	11·17	6·25	4·44	72·72	3252	1746	2087	0·2722	305	999
68	5650	771	11·07	6·21	4·41	71·18	3162	1712	2020	0·2706	283	1000
69	5700	721	10·97	6·17	4·38	69·71	3075	1677	1957	0·2693	261	1000
70		671	10·88	6·12		68·39	2996	1642	1901	0·2683	239	1001

Table 1.3 (cont.)

I	Radius (km)	Depth (km)	V_P (km/sec)	V_S (km/sec)	ρ (g/cm³)	ϕ (km²/sec²)	k (kb)	μ (kb)	λ (kb)	σ	p (kb)	g (cm/sec²)
71	5700	671	10·08	5·21	4·05	65·37	2644	1098	1911	0·3175	239	1001
72	5725	646	10·01	5·21	4·02	64·03	2575	1092	1846	0·3141	228	1001
73	5750	621	9·95	5·22	4·00	62·62	2503	1088	1777	0·3100	218	1000
74	5775	596	9·88	5·23	3·97	61·16	2430	1086	1706	0·3055	208	1000
75	5800	571	9·81	5·24	3·95	59·66	2355	1084	1632	0·3004	199	1000
76	5825	546	9·75	5·26	3·92	58·13	2280	1084	1558	0·2948	189	999
77	5850	521	9·68	5·27	3·90	56·58	2205	1084	1482	0·2887	179	999
78	5875	496	9·61	5·29	3·87	55·05	2131	1084	1408	0·2825	169	999
79	5900	471	9·54	5·31	3·85	53·56	2060	1083	1338	0·2763	160	998
80	5925	446	9·48	5·32	3·82	52·14	1992	1079	1272	0·2705	150	998
81	5951	420	9·41	5·32	3·80	50·76	1926	1074	1210	0·2649	140	997
82	5951	420	8·75	4·67	3·58	47·61	1706	780	1186	0·3017	140	997
83	5975	396	8·71	4·66	3·57	46·79	1668	775	1151	0·2988	132	997
84	6000	371	8·66	4·66	3·54	46·02	1630	768	1118	0·2965	123	996
85	6050	321	8·56	4·60	3·49	44·96	1569	739	1076	0·2964	105	994
86	6100	271	8·46	4·50	3·44	44·53	1530	695	1067	0·3028	88	992
87	6150	221	8·35	4·37	3·39	44·26	1498	648	1066	0·3110	71	990
88	6175	196	8·30	4·34	3·37	43·79	1473	633	1051	0·3119	63	989
89	6200	171	8·25	4·35	3·35	42·77	1432	633	1009	0·3072	54	988
90	6225	146	8·19	4·37	3·34	41·60	1388	637	962	0·3008	46	987
91	6250	121	8·13	4·44	3·33	39·88	1327	565	889	0·2878	38	986
92	6271	100	8·08	4·53	3·32	38·05	1264	681	810	0·2717	31	986
93	6271	100	8·08	4·53	3·32	38·05	1264	681	810	0·2717	31	986
94	6290	81	8·04	4·62	3·32	36·21	1202	708	730	0·2537	25	985
95	6310	61	8·00	4·72	3·32	34·24	1135	738	643	0·2327	18	984
96	6330	41	7·95	4·80	3·31	32·52	1076	762	568	0·2136	12	984
97	6350	21	7·91	4·83	3·30	31·45	1037	769	524	0·2026	5	983
98	6350	21	6·20	3·40	2·79	23·03	642	322	427	0·2850	5	983
99	6360	11	6·20	3·40	2·79	23·03	642	322	427	0·2850	3	982
100	6371	0	6·20	3·40	2·79	23·03	642	322	427	0·2850	0	981

excited by the Columbian earthquake to be observed. The eigenperiods reported in their two papers constitute slightly more than 200 gross Earth data (GED), of which more than half are overtones. Some of the data are redundant. Gilbert (1971) has shown how to construct a non-redundant sub-set, a significant Earth datum (SED), and Gilbert *et al.* (1973) showed that the number of SED contained in the normal mode data in the two papers of Dziewonski and Gilbert (1972, 1973a) is 38. In a later paper (in preparation) Dziewonski and Gilbert identified over 400 additional overtones. The number of SED in the data set of over 600 GED is only 46. Thus a 200 per cent increase in the number of GED results in only a 20 per cent increase in the number of SED. It appears now that the quality, rather than the quantity, of GED is of equal importance. Dziewonski and Gilbert used this additional normal mode data to construct two Earth models of an average, radially symmetric Earth. Details of these models are given in their 1973 paper.

The new data narrow the range of uncertainty in the density distribution and contain more information on the velocity distribution than the published data on the travel-times of body waves. Because of the large wavelengths of the normal modes, their eigenfrequencies are likely to be more closely related to the properties of a spherically symmetric, radially stratified "average" Earth than the nearly exclusive land-based travel-time observations. The resolving power of the normal mode data set shows that the average velocities and densities are known in the mantle to one per cent for averaging lengths of

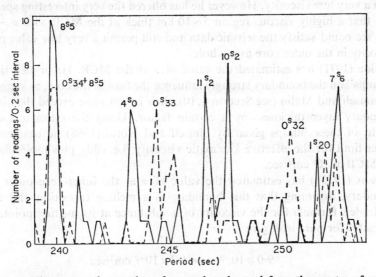

Fig. 1.16. Histograms of a number of spectral peaks read from the spectra of vertical component readings. The histogram for seismograms with a lag time of less than 18 hr, broken line; over 18 hr, solid line. (After Dziewonski and Gilbert, 1973a.)

~200 km (Gilbert *et al.*, 1973). For the same precision the averaging lengths in the outer core are ~400 km.

Dziewonski and Gilbert (1973a) also observed an inner core mode ($_{11}S_2$) in 11 spectra of seismograms with large lag times—over two days after the origin time in some cases (see Fig. 1.16). This represents direct proof of the solidity of the inner core—the mode $_{11}S_2$ could not exist if the inner core were liquid. The inner core mode $_{11}S_2$ is coupled with a compressional mode of nearly identical phase velocity ($_{10}S_2$)—the observed periods are 246·89 and 247·74 sec respectively. The coincidence of $_{10}S_2$ and $_{11}S_2$ implies that the average shear velocity in the inner core must be ~3·6 km/sec, and that dissipation in the inner core is low.

1.7. Viscosity of the Earth's Core

The viscosity of the Earth's core is one of the least well known physical parameters of the Earth. Estimates in the literature of the kinematic viscosity ν differ by many orders of magnitude, ranging from 10^{-3} to 10^{11} cm^2/sec. High values of the viscosity have been deduced from the observation that compressional waves traverse the core without suffering appreciable attenuation (see e.g. Rochester, 1970). Such results lead to a value of ν of the order of 10^9–10^{11} cm^2/sec. Gans (1972a) has shown that the effect of the Earth's magnetic field at the MCB cannot reconcile the seismic data for *SH* waves with a very low viscosity. However he has offered the very interesting speculation that a highly viscous region $5 \sim 10$ km thick at the MCB with $\nu \sim 10^{11}$ cm^2/sec could satisfy the seismic data and still permit a very low value of the viscosity in the outer core as a whole.

Hide (1971) has estimated the value of ν at the MCB. He argued that if "bumps" on the boundary strongly influence the flow in the core, as suggested by himself and Malin (see Section 4.10), their height must exceed the viscous boundary layer thickness by a certain factor. Using the estimate for the height of these bumps given by himself and Horai (1968) he obtained an upper limit for the effective kinematic viscosity (i.e. eddy plus molecular) at the MCB of 10^6 cm^2/sec.

Gans (1972b) has estimated the value of ν at the inner core–outer core boundary, assuming that this boundary is a melting transition and using Andrade's formula for the viscosity of a substance at its melting point. The estimates for pure iron give

$$9\cdot0 \times 10^{-3} < \nu < 1\cdot7 \times 10^{-2} \text{ cm}^2/\text{sec}$$

The core, however, is not pure iron, but contains some 10 per cent lighter alloying element—both silicon and sulphur have been proposed (see Section

5.5). Both substances would lower the liquidus temperature, density and average atomic weight. The experimental data on the effect of alloying on the viscosity however are very sparse, although it seems probable that it would be lowered. It is not possible to estimate with any certainty the amount of the decrease. Gans suggests that in the core

$$2 \cdot 8 \times 10^{-3} < \nu < 1 \cdot 5 \times 10^{-2} \text{ cm}^2/\text{sec}$$

with a typical value of 6×10^{-3} cm²/sec.

This range of values applies at the boundary of the inner and outer core. However if the melting point gradient is very shallow throughout the core (see Higgins and Kennedy, 1971 and Section 3.3), this range will not change much. With these values, the precessional mechanism proposed by Malkus to generate the Earth's magnetic field (see Section 4.7) would be unstable. Whether it would be sufficiently unstable to overcome the highly sub-adiabatic temperature gradient in the core as proposed by Higgins and Kennedy cannot be determined.

The free gravitational oscillations of the Earth's inner core and its probable detection in the gravimetric data for the Chilean earthquake of May 1960 were first considered by Slichter (1961). More recently Won and Kuo (1973) have extended Slichter's work by considering the effects of rotation, viscous dissipation and magnetic stresses. They suggested that oscillations of the inner core may be excited by an external agency such as a large earthquake or the impact of a large meteorite. They found that the period of oscillation of the inner core is virtually independent of the viscosity and is of the order of several hours, considerably longer than that obtained by Slichter. The decay time of the oscillations is determined mainly by the viscosity of the outer core at the inner core boundary and is very long—of the order of thousands of years. Because of the rotation of the Earth, the inner core, when it is oscillating, also precesses with a period of about 36 hr, dependent mainly on the density contrast between the inner and outer cores.

Won and Kuo estimated that the amplitude of the oscillation of the inner core is about 60 cm if the available energy is 4×10^{21} erg from a 10^{25} erg earthquake (magnitude 8.5). If only 2×10^{21} erg is available, the amplitude is about 45 cm and about 30 cm if only 1×10^{21} erg is available. The maximum observable surface gravity anomalies for these three cases are 0.074, 0.058 and 0.038 μgal respectively which are near or below the noise level of tidal gravity measurements at favourable sites.

Although the amplitude of the oscillations of the inner core is extremely small, the decay time is long and Won and Kuo suggested that it may contribute to the setting up of the velocity field in the outer core necessary for the geomagnetic dynamo (see Section 4.2). If the amplitude of the oscillations is about 50 cm and if the period is 7·4 hr, the velocity at the surface of the inner core is ≃0·01 cm/sec radially, in agreement with the estimate of the radial velocity given by Bullard and Gellman (1954). The disturbance may be slowly propagated through the outer core if the motion lasts sufficiently long. Again the intensity of the Earth's magnetic field is decreasing currently at about 5 per cent/century. Assuming an exponential decay of the field this is equivalent to a time constant of about 2000 yr. The viscosity of the outer core corresponding to a decay time constant of about 2000 yr is of the order of 0·1 poise which is very close to the value suggested by Gans (1972b).

Although Won and Kuo included hydromagnetic forces in their equation of motion, they later dismiss them as being negligible. However Busse (1974) has pointed out that, following Toomre's (1966) discussion of the analogous problem of core-mantle coupling during the 26,000 yr precession of the Earth, magnetic stresses will exceed viscous stresses by several orders of magnitude if a viscosity of 1 poise or less is assumed (as is favoured by Won and Kuo). This would considerably shorten the decay time by ohmic dissipation. Although magnetic and viscous stresses are solely responsible for determining the decay rates of oscillations they have but little effect on their periods. However Busse (1974) has shown that the finite radius of the outer core and the action of the Coriolis force on the motion of the fluid in the outer core, could change the period by as much as 50 per cent although these two effects act in opposite directions. For a density contrast of 0·3 g/cm³ between the inner and outer core, Busse obtained a period of 7·3 hr which is only 5 per cent different from the value of Won and Kuo, since the effects of rotation and of a finite outer core radius nearly compensate each other at this period.

Crossley and Smylie (1975) have considered in detail the rate at which energy is dissipated in the liquid outer core of the Earth. There are two possible dissipative mechanisms, ohmic and viscous, and they showed that oscillations in the core are virtually unaffected by either damping mechanism. Most of the ohmic dissipation takes place within a few "skin-depths" of the boundaries—Crossley and Smylie expect that a similar result would be obtained for viscous dissipation, although their calculations were confined to estimating laminar dissipation in the body of the liquid core. The conclusion that the Earth's core appears to be nearly free of dissipation for small harmonic oscillations such as those which occur in the natural elastic oscillations of the Earth or those which may occur as gravitational oscillations if the temperature profile in the core is subadiabatic (see Section 1.3 and 3.5) suggests that modes of oscillation which are largely confined to the core may persist for unexpectedly long times—perhaps long enough to be important in driving the geodyanmo (see Section 4.7).

References

Adams, R. D. (1972). Multiple inner core reflections from a Novaya Zemlya explosion. *Bull. seism. Soc. Am.* **62**, 1063.

Adams, R. D. and Randall, M. J. (1963). Observed triplication of PKP. *Nature* **200**, 744.

Adams, R. D. and Randall, M. J. (1964). The fine structure of the Earth's core. *Bull. seism. Soc. Am.* **54**, 1299.

Alterman, Z., Jarosch H. and Pekeris, C. L. (1959). Oscillations of the Earth. *Proc. R. Soc.* A **252**, 80.

Anderson, D. L. (1967). A seismic equation of state. *Geophys. J.* **13**, 9.

Anderson, D. L. and Hanks, T. C. (1972). Formation of the Earth's core. *Nature* **237**, 387.

Anderson, D. L., Sammis, C. and Jordan, T. H. (1970). Composition and evolution of the mantle and the core. *Science* **171**, 1103.

Anderssen, R. S. and Seneta, E. (1971). A simple statistical estimation procedure for Monte Carlo inversion in geophysics. *Pure appl. Geophys.* **91**, 5014.

Anderssen, R. S. and Seneta, E. (1972). A simple statistical estimation procedure for Monte Carlo inversion in geophysics. II: Efficiency and Hempel's praadox. *Pure appl. Geophys.* **96**, 5.

Anderssen, R. S., Worthington, M. H. and Cleary, J. R. (1972). Density modelling by Monte Carlo inversion. I: Methodology. *Geophys. J.* **29**, 433.

Backus, G. (1971). Inference from inadequate and inaccurate data. *In* "Mathematical Problems in the Geophysical Sciences" (Ed. W. H. Reid), Lectures in Applied Mathematics, Vol. 14, *Am. math. Soc.*

Backus, G. E. and Gilbert, J. F. (1961). The rotational splitting of the free oscillations of the Earth. *Proc. natn. Acad. Sci. U.S.A.* **47**, 362.

Backus, G. E. and Gilbert, J. F. (1967). Numerical applications of a formalism for geophysical inverse problems. *Geophys. J.* **13**, 247.

Backus, G. E. and Gilbert, J. F. (1968). The resolving power of gross Earth data. *Geophys. J.* **16**, 169.

Backus, G. and Gilbert, F. (1970). Uniqueness in the inversion of inaccurate gross Earth data. *Phil. Trans. R. Soc.* A **266**, 123.

Balakina, L. M. and Vvedenskaya, A. V. (1962). *Izv. Akad. Nauk SSR Ser. Geofiz* **11**, 909.

Bertrand, A. E. S. and Clowes, R. M. (1974). Seismic array evidence for a two-layer core transition zone. *Phys. Earth Planet. Int.* **8**, 251.

Berzon, I. S., Kogan, S. D. and Pessechnik, I. P. (1972). The character of the mantle-core boundary from observations of PCP waves. *Earth Planet. Sci. Letters* **16**, 166.

Birch, F. (1952). Elasticity and constitution of the Earth's interior. *J. geophys. Res.* **57**, 227.

Birch, F. (1961). The velocity of compressional waves in rocks to 10 kilobars, 2. *J. geophys. Res.* **66**, 2199.

Birch, F. (1964). Density and composition of mantle and core. *J. geophys. Res.* **69**, 4377.

Bolt, B. A. (1959). Travel times of PKP up to 145°. *Geophys. J.* **2**, 190.

Bolt, B. A. (1962). Gutenberg's early *PKP* observations. *Nature*, **196**, 122.

Bolt, B. A. (1964). The velocity of seismic waves near the Earth's centre. *Bull. seism. Soc. Am.* **54**, 191.

Bolt, B. A. (1970). *PdP* and *PKiKP* waves and diffracted *PcP* waves. *Geophys. J.* **20**, 367.

Bolt, B. A. (1972). The density distribution near the base of the mantle and near the Earth's centre. *Phys. Earth Planet. Int.* **5**, 301.

Bolt, B. A. and Qamar, A. (1970). Upper bound to the density jump at the boundary of the Earth's inner core. *Nature*, **228**, 148.

Bolt, B. A., O'Neill M. and Qamar, A. (1968). Seismic waves near 110°: is structure in core or upper mantle responsible? *Geophys. J.* **16**, 475.

Bolt, B. A., Niazi, M. and Somerville, M. (1970). Diffracted *ScS* and the shear velocity of the core boundary. *Geophys. J.* **19**, 299.

Buchbinder, G. G. R. (1968). Properties of the core-mantle boundary and observations of *PcP*. *J. geophys. Res.* **73**, 5901.

Buchbinder, G. G. R. (1971). A velocity structure of the Earth's core. *Bull. seism. Soc. Am.* **61**, 429.

Buchbinder, G. G. R. and Poupinet, G. (1973). Problems related to *PcP* and the core–mantle boundary illustrated by two nuclear events. *Bull. seism. Soc. Am.* **63**, 2047.

Bullard, E. C. and Gellman, H. (1954). Homogeneous dynamos and terrestrial magnetism. *Phil. Trans. R. Soc.* A **247**, 213.

Bullen, K. E. (1954). "Seismology", Methuen and Co. Ltd., London.

Bullen, K. E. (1963). "An Introduction to the Theory of Seismology", Cambridge University Press, London.

Bullen, K. E. (1965). Models for the density and elasticity of the Earth's lower core. *Geophys. J.* **9**, 233.

Bullen, K. E. and Haddon, R. A. W. (1970). Evidence from seismology and related sources on the Earth's present internal structure. *Phys. Earth Planet. Int.* **2**, 342.

Busse, F. H. (1974). On the free oscillation of the Earth's inner core. *J. geophys. Res* **79**, 753.

Caloi, P. (1961). Seismic waves from the outer core and the inner core. *Geophys. J.* **4**, 139.

Chowdhury, D. K. and Frasier, C. W. (1973). Observations of *PcP* and *P* phases at LASA at distances from 26° to 40°. *J. geophys. Res.* **78**, 6021.

Chung, D. H. (1974). General relationship among sound speeds. I: New experimental information. *Phys. Earth Planet. Int.* **8**, 113.

Clark, S. P., Jr. and Ringwood, A. E. (1964). Density distribution and constitution of the mantle. *Rev. Geophys.* **2**, 35.

Cleary, J. (1969). The *S*-velocity at the core–mantle boundary from observations of diffracted *S*. *Bull. seism. Soc. Am.* **59**, 1399.

Cleary, J. R. (1974). The *D″* region. *Phys. Earth Planet. Int.* **9**, 13.

Cleary, J. R. and Haddon, R. A. W. (1972). Seismic wave scattering near the core–mantle boundary: a new interpretation of precursors to *PKP*. *Nature*, **240**, 549.

Cook, A. H. (1963). The contribution of observations of satellites to the determination of the Earth's gravitational potential. *Space Sci. Rev.* **2**, 355.

Crossley, D. J. and Smylie, D. E. (1975). Electromagnetic and viscous damping of core oscillations. *Geophys. J.* (in press).

Denson, M. E. (1952). Longitudinal waves through the Earth's core. *Bull. seism. Soc. Am.* **42**, 119.

Derr, J. S. (1969). Internal structure of the Earth inferred from free oscillations. *J. geophys. Res.* **74**, 5202.

Doornbos, D. J. (1974). The anelasticity of the inner core. *Geophys. J.* **38**, 397.

Dorman, J., Ewing, J. and Alsop, L. (1965). Oscillations of the Earth: new core-mantle boundary model based on low-order free vibrations. *Proc. natn. Acad. Sci., U.S.A.* **54**, 364.

Dziewonski, A. M. (1970). Correlation properties of free period partial derivatives and their relation to the resolution of gross Earth data. *Bull. seism. Soc. Am.* **60**, 741.

Dziewonski, A. M. (1971). Overtones of free oscillations and the structure of the Earth's interior. *Science* **172**, 1336.

Dziewonski, A. M. and Gilbert, F. (1971). Solidity of the inner core of the Earth inferred from normal mode observations. *Nature*, **234**, 465.

Dziewonski, A. M. and Gilbert, F. (1972). Observations of normal modes from 84 recordings of the Alaska earthquake of 1964, March 28. *Geophys. J.* **27**, 393.

Dziewonski, A. M. and Gilbert, F. (1973a). Observations of normal modes from 84 recordings of the Alaska earthquake of 1964, March 28. II: Further remarks based on new spheroidal overtone data. *Geophys. J.* **35**, 401.

Dziewonski, A. M. and Gilbert, F. (1973b). Identification of normal modes using spectral stacking and stripping. *Trans. Am. geophys. Un.* **54**, 374.

Dziewonski, A. M. and Haddon, R. A. W. (1974). The radius of the core–mantle boundary inferred from travel-time and free oscillation data; a critical review. *Phys. Earth Planet. Int.* **9**, 28.

Engdahl, E. R. (1968). Core phases and the Earth's core. Ph.D. thesis, St. Louis University.

Engdahl, E. R. and Johnson, L. E. (1972). A new *PcP* data set from nuclear explosions on Amchitka Island. *Trans. Am. geophys. Un.* **53**, 1045.

Engdahl, E. R., Flinn, E. A. and Romney, C. F. (1970). Seismic waves reflected from the Earth's inner core. *Nature*, **228**, 852.

Ergin, K. (1967). Seismic evidence for a new layered structure of the Earth's core. *J. geophys. Res.* **72**, 3669.

Franklin, J. N. (1970). Well-posed stochastic extensions of ill-posed linear problems. *J. math. Analysis Applic.* **31**, 682.

Gans, R. F. (1972a). Reflection of *SH* in the presence of a magnetic field. *Geophys J.* **29**, 173.

Gans, R. F. (1972b). Viscosity of the Earth's core. *J. geophys. Res.* **77**, 360.

Gilbert, F. (1971). Ranking and winnowing gross Earth data for inversion and resolution. *Geophys. J.* **23**, 125.

Gilbert, F. and Helmberger, D. (1972). Generalized ray theory for a layered sphere. *Geophys. J.* **27**, 57.

Gilbert, F., Dziewonski, A. M. and Brune, J. (1973). An informative solution to a seismological inverse problem. *Proc. natn. Acad. Sci. U.S.A.* **70**, 1410.

Gutenberg, B. (1913). Über die Konstitution des Erdinnern, erschlossen aus Erdbebenbeobachtungen. *Phys. Z.* **14**, 1217.

Gutenberg, B. (1957). The "boundary" of the Earth's inner core. *Trans. Am. geophys. Un.* **38**, 750.

Gutenberg, B. (1958a). Wave velocities in the Earth's core. *Bull. seism. Soc. Am.* **48**, 301.

Gutenberg, B. (1958b). Caustics produced by waves through the Earth's core. *Geophys. J.* **1**, 238.

Gutenberg, B. (1959). "Physics of the Earth's Interior", Academic Press, New York.

Gutenberg, B. and Richter, C. F. (1938). *P* and the Earth's core. *Mon. Not. R. astr. Soc. geophys. Suppl.* **4**, 363.

Gutenberg, B. and Richter, C. F. (1939). On seismic waves. *Gerl. Beit. Geophys.* **54**, 94.

Haddon, R. A. W., (1972). Corrugations on the mantle–core boundary or transition layers between inner and outer core? *Trans. Am. geophys. Un.* **53**, 600.

Haddon, R. A. W. and Bullen, K. E. (1969). An Earth model incorporating free Earth oscillation data. *Phys. Earth Planet. Int.* **2**, 35.

Haddon, R. A. W. and Cleary, J. R. (1974). Evidence for scattering of seismic *PKP* waves near the mantle–core boundary. *Phys. Earth Planet. Int.* **8**, 211.

Hai, N. (1961). Propagation des ondes longitudinales dans le noyeau terrestre d'après les seismes profonds de Iles Fidji. *Annls Géophys.* **17**, 60.

Hai, N. (1963). Propagation des indes longitudinales dans le noyeau terrestre. *Annls Géophys.* **19**, 285.

Hales, A. L. and Roberts, J. L. (1970a). The travel times of *S* and *SKS*. *Bull. seism. Soc. Am.* **60**, 461.

Hales, A. L. and Roberts, J. L. (1970b). Shear velocities in the lower mantle and the radius of the core. *Bull. seism. Soc. Am.* **60**, 1427.

Hales, A. L. and Roberts, J. L. (1971). The velocities in the outer core. *Bull. seism. Soc. Am.* **61**, 1051.

Hannon, W. J. and Kovach, R. L. (1966). Velocity filtering of seismic core phases. *Bull. seism. Soc. Am.* **54**, 441.

Herrin, E. (1968). Introduction to the 1968 Seismological Tables for *P* phases. *Bull. seism. Soc. Am.* **58**, 1193.

Hide, R. (1971). Viscosity of the Earth's core. *Nature Phys. Sci.* **233**, 100.

Hide, R. and Horai, K. I. (1968). On the topography of the core–mantle interface. *Phys. Earth Planet. Int.* **1**, 305.

Higgins, G. and Kennedy, G. C. (1971). The adiabatic gradient and the melting point gradient in the core of the Earth. *J. geophys. Res.* **76**, 1870.

Ibrahim, A. K. (1971). The amplitude ratio *PcP/P* and the core–mantle boundary. *Pure appl. Geophys.* **91**, 114.

Ibrahim, A. K. (1973). Evidences for a low velocity core–mantle transition zone. *Phys. Earth Planet. Int.* **7**, 187.

Jackson, D. D. (1972). Interpretation of inaccurate, insufficient and inconsistent data. *Geophys. J.* **28**, 97.

Jackson, D. D. and Anderson, D. L. (1970). Physical mechanisms of seismic-wave attenuation. *Rev. Geophys. Space Phys.* **8**, 1.

Jacobs, J. A. (1968). The structure of the Earth's core. *Phys. Earth Planet. Int.* **1**, 196.

Jacobs, J. A. (1970). Geophysical Numerology. *Nature*, **227**, 161.

Jeffreys, H. (1936). On travel times in seismology. *Bur. Cent. Seism. Inter. A., Fasc* **14**, 1.

Jeffreys, H. (1937). On the materials and density of the Earth's crust. *Mon. Not. R. astr. Soc. geophys. Suppl.* **4**, 50.

Jeffreys, H. (1939a). The times of *P, S* and *SKS* and the velocities of *P* and *S*. *Mon. Not. R. astr. Soc. geophys. Suppl.* **4**, 498.

Jeffreys, H. (1939b). The times of *PcP* and *ScS*. *Mon. Not. R. astr. Soc. geophys. Suppl.* **4**, 537.

Jeffreys, H. (1939c). The times of the core waves. *Mon. Not. R. astr. Soc. geophys Suppl.* **4**, 548.

Jeffreys, H. (1939d). The times of the core waves. *Mon. Not. R. astr. Soc. geophys. Suppl.* **4**, 594.

Jeffreys, H. (1961). "Theory of Probability", Clarendon Press, Oxford.

Jeffreys, H. and Bullen, K. E. (1935). Times of transmission of earthquake waves. *Bur. Cent. Seism. Inter. A. Fasc.* **11**.

Jeffreys, H. and Bullen, K. E. (1940). "Seismological Tables", Brit. Assoc. Gray-Milne Trust.

Johnson, L. R. (1967). Array measurements of *P* velocities in the upper mantle. *J. geophys. Res.* **72**, 6309.

Johnson, L. R. (1969). Array measurements of *P* velocities in the lower mantle. *Bull. seism. Soc. Am.* **59**, 973.

Jordan, T. H. (1972). Estimation of the radial variation of seismic velocities and density in the Earth. Ph.D. thesis, California Institute of Technology.

Jordan, T. H. and Anderson, D. L. (1974). Earth structure from free oscillations and travel times. *Geophys. J.* **36**, 411.

Julian, B. R. and Sengupta, M. K. (1973). Seismic travel-time evidence for lateral heterogeneity in the deep mantle. *Nature*, **242**, 443.

Julian, B. R., Davies, D. and Sheppard, R. M. (1972). PKJKP. *Nature*, **235**, 317.

Keilis-Borok, V. I. and Yanovskaya, T. R. (1967). Inverse problems of seismology. *Geophys. J.* **13**, 223.

King, D. W., Haddon, R. A. W. and Cleary, J. R. (1973). Evidence for seismic wave scattering in the "D" layer. *Earth Planet. Sci. Letters* **20**, 353.

King, D. W., Haddon, R. A. W. and Cleary, J. R. (1974). Array analysis of precursors to *PKIKP* in the distance range 128° to 142°. *Geophys. J.* **37**, 157.

Knopoff, L. (1964). *Q. Rev. Geophys.* **2**, 625.

Knopoff, L. and MacDonald, G. J. F. (1958). The magnetic field and the central core of the Earth. *Geophys. J.* **1**, 216.

Kogan, S. D. (1972). A study of the dynamics of a longitudinal wave reflected from the Earth's core. *Bull. Acad. Sci. USSR Earth Phys.* **6**, 3.

Landisman, M., Sato, Y. and Nafe, J. (1965). Free vibrations of the Earth and the properties of its deep interior regions. 1: Density. *Geophys. J.* **9**, 439.

Lehmann, I. (1936). "P". *Publ. Bur. Cent. Seism. Int. Ser. A* **14**, 3.

McMechan, G. A. and Wiggins, R. A. (1972). Depth limits in body wave inversions. *Geophys. J.* **28**, 459.

Mitchell, B. J. and Helmberger, D. V. (1973). Shear velocities at the base of the mantle from observations of *S* and *ScS. J. geophys. Res.* **78**, 6009.

Mizutani, H. and Abe, K. (1972). An Earth model consistent with free oscillation and surface wave data. *Phys. Earth Planet. Int.* **5**, 345.

Moore, E. H. (1920). *Bull. Am. math. Soc.* **26**, 394.

Müller, G. (1973). Amplitude studies of core phases. *J. geophys. Res.* **78**, 3469.

Nuttli, O. W. (1969). Travel times and amplitudes of *S* waves from nuclear explosions in Nevada. *Bull. seism. Soc. Am.* **59**, 385.

Parker, R. L. (1970). The inverse problem of electrical conductivity in the mantle. *Geophys. J.* **22**, 121.

Parker, R. L. (1972a). The Backus–Gilbert method and its application to the electrical conductivity problem. *In* "Mathematics of Profile Inversion" (Ed. L. Colin), *NASA Tech. Mem.* **62**, 150.

Parker, R. L. (1972b). Inverse theory with grossly inadequate data. *Geophys. J.* **29**, 123.

Pekeris, C. L. (1966). The internal constitution of the Earth. *Geophys. J.* **11**, 85.

Pekeris, C. L., Alterman, Z. and Jarosch, H. (1961). Rotational multiplets in the spectrum of the Earth. *Phys. Rev.* **122**, 1692.

Pekeris, C. L. and Accad, Y. (1972). Dynamics of the liquid core of the Earth. *Phil. Trans. R. Soc. A* **273**, 237.

Penrose, R. (1955). A generalized inverse for matrices. *Proc. Camb. phil. Soc.* **51**, 406.

Press, F. (1968a). Density distribution in Earth. *Science* **160**, 1218.

Press, F. (1968b). Earth models obtained by Monte Carlo inversion. *J. geophys. Res.* **73**, 5223.

Press, F. (1970a). Earth models consistent with geophysical data. *Phys. Earth Planet. Int.* **3**, 3.

Press, F. (1970b). Regionalized Earth models. *J. geophys. Res.* **75**, 6575.

Qamar, A. (1973). Revised velocities in the Earth's core. *Bull. seism. Soc. Am.* **63**, 1073.

Qamar, A. and Eisenberg, A. (1974). The damping of core waves. *J. geophys. Res.* **79**, 758.

Randall, M. J. (1970). *SKS* and seismic velocities in the outer core. *Geophys. J.* **21**, 441.

Ringwood, A. E. (1972). Mineralogy of the deep mantle: current status and future developments. *In* "The Nature of the Solid Earth" (Ed. E. C. Robertson), McGraw-Hill, New York.

Robinson, R. and Kovach, R. (1972). Shear wave velocities in the Earth's mantle. *Phys. Earth Planet. Int.* **5**, 30.

Rochester, M. G. (1970). Core–mantle interactions: geophysical and astronomical consequences. *In* "Earthquake Displacement Fields and the Rotation of the

Earth" (Ed. L. Mansinha, D. E. Smylie and A. E. Beck), D. Reidel Publ. Co., Holland.

Sacks, I. S. (1971). Anelasticity of the outer core. *Carnegie Inst. Yearbook* **69**, 414.

Sato, R. and Espinosa, A. F. (1967). Dissipation in the Earth's mantle and rigidity and viscosity in the Earth's core determined from waves multiply reflected from the mantle–core boundary. *Bull. seism. Soc. Am.* **57**, 829.

Slichter, L. B. (1961). Fundamental free mode of the Earth's inner core. *Proc. natn. Acad. Sci. U.S.A.* **47**, 186.

Shankland, T. J. and Chung, D. H. (1974). General relationships among sound speeds. II: Theory and discussion. *Phys. Earth Planet. Int.* **8**, 121.

Simmons, G. and Chung, D. H. (1968). A powder method for determining the elastic parameters of a solid. *Trans. Am. geophys. Un.* **49**, 308.

Smylie, D. E. (1973). Dynamics of the outer core. *Proc. 2nd Int. Symp. Geod. Phys. Earth, Potsdam.*

Smylie, D. E. and Mansinha, L. (1971). The elasticity theory of dislocation in real Earth models and changes in the rotation of the Earth. *Geophys. J.* **23**, 329.

Subiza, G. P. and Bäth, M. (1964). Core phases and the inner core boundary. *Geophys. J.* **8**, 496.

Suzuki, Y. and Sato, R. (1970). Viscosity determination in the Earth's outer core from *ScS* and *SKS* phases. *J. Phys. Earth* **18**, 157.

Taggart, J. and Engdahl, E. R. (1968). Estimation of *PcP* travel times and depth to the core. *Bull. seism. Soc. Am.* **58**, 1293.

Toomre, A. (1966). On the coupling of the Earth's core and mantle during the 26,000 year precession. *In* "The Earth–Moon System" (Ed. B. G. Marsden and A. G. Cameron), Plenum Press, New York.

Wang, C.-Y. (1970). Density and constitution of the mantle. *J. geophys. Res.* **75**, 3264.

Wang, C.-Y. (1972). A simple Earth model. *J. geophys. Res.* **77**, 4318.

Wiggins, R. A. (1969). Monte Carlo inversions of body wave observations. *J. geophys. Res.* **74**, 3171.

Wiggins, R. A. (1972). The general linear inverse problem: implication of surface waves and free oscillations for Earth structure. *Rev. geophys. Space Phys.* **10**, 251.

Wiggins, R. A., McMechan, G. A. and Toksöz, M. N. (1973). Range of Earth structure nonuniqueness implied by body wave observations. *Rev. geophys. Space Phys.* **11**, 87.

Williamson, E. D. and Adams, L. H. (1923). Density distribution in the Earth. *J. Wash. Acad. Sci.* **13**, 413.

Won, I. J. and Kuo, J. T. (1973). Oscillation of the Earth's inner core and its relation to the generation of geomagnetic field. *J. geophys. Res.* **78**, 905.

Worthington, M. H., Cleary, J. R. and Anderssen, R. S. (1972). Density modelling by Monte Carlo inversion. II: Comparison of recent Earth models. *Geophys. J.* **29**, 445.

Wunsch, C. (1974). Simple models of the deformation of an Earth with a fluid core, I. *Geophys. J.* **39**, 413.

Wunsch, C. (1975). Simple models of the deformation of an Earth with a fluid core, II. Dissipation and magnetohydrodynamic effects. *Geophys. J.* **41**, 165.

2. The Origin of the Core

2.1. Introduction

A fundamental question in any discussion of the Earth's core is its origin. Has the Earth always had a core, or has its present structure evolved over geologic time? Such a question cannot be divorced from the much broader issue of the origin of the Earth itself. There is one constraint however that can be imposed upon possible evolutions. Most theories of the origin of the Earth's magnetic field ascribe it to motions in the liquid outer core (see Section 4.2). Rocks as old as 2700 m yr have been found which possess remanent magnetization, so that it is extremely probable that the Earth had a molten outer core, comparable in size to that at present at least that long ago. No similar deductions can be made about the state of the inner core.

With regard to the broader question of the origin and evolution of the solar system, it must not be forgotten that it is by no means certain *a priori* that the problem can be given a definite answer. It may well be that all memory has been lost of the circumstances under which our solar system was born. All that we can do is to attempt to derive its present state from an assumed event or series of events which occurred in the distant past. Thus in a sense the method is one of trial and error. It is more reasonable to assume that planets are normally present in the vicinity of certain stars than to suppose that our planetary system is unique or at least very rare—there are about 200,000 million stars in our galaxy alone. The origin of the solar system is thus part of a very much larger problem—the evolution of the sun and stars in general. A symposium on the origin of the solar system was held recently at Nice and the proceedings (Reeves, 1972) provide a good summary of current ideas.

Theories of the origin of the Earth and other members of the solar system may be classified in a number of ways. The Earth may have cooled from a hot gas or accreted cold from dust particles, probably of composition similar to that of chondritic meteorites. In the past most theories of the origin of the Earth have assumed that the proto-Earth was homogeneous and that the present differentiation into a core, mantle and crust occurred later. There are, however, a number of difficulties with such theories, and there has been some discussion in recent years of non-homogeneous models. In such models, the core of the Earth is formed first and the mantle deposited upon it later. Non-homogeneous models can also have either a hot or a cold origin. A hot origin involves a consideration of the order of condensation of the elements in the primitive solar nebula; a cold origin involves a study of the mechanics of the accretion process.

2.2. The Accretion Mechanism

For the smallest particle sizes, accumulation cannot be very rapid. The particles will collide, partly as a result of their Brownian motion in the gas, and partly as a result of acceleration by very weak electric fields which can be expected to be produced in the nebula through the interaction of convection with ionization produced by natural radioactivity. This state of accumulation may be greatly accelerated as a result of the natural magnetism of the interstellar grains. Purcell and Spitzer (1971) believe that the interstellar grains most probably would be ferromagnetic or super-paramagnetic so that they could be aligned by the very weak interstellar magnetic field. If this is the case, and it appears to be consistent with evidence from meteorites (Brecher, 1971), then the interstellar grains would be able to come together to form considerably larger units, both during the late stages of the collapse that formed the primitive solar nebula and during the early history of the solar nebula itself. The cross section for magnetic capture of one particle by another would be very much greater than the geometric cross sections involved in ordinary non-magnetic collisions.

The accretion mechanism is one of the most difficult of all problems connected with the origin of the solar system. Reeves (1972) was moved to write "some of the texts written on the subject carry a definite resemblance to the writing of the medieval alchemists". At least two stages may be distinguished in the evolution of the solar nebula and formation of the planets: the first when solid particles were small and their chaotic velocities were damped by friction in the gas, causing them to collect rapidly towards the central plane of the nebula where asteroid-sized bodies were formed; the second when mutual gravitational perturbations of asteroid-sized bodies increased eccentricities and inclinations of their orbits thus again thickening the volume of space occupied by solid material. This second stage is likely to have been much longer.

The initial step is the presence of interstellar grains in the gas, which after some degree of gas absorption, find themselves concentrated in the nebular plane and somehow manage to accrete into large bodies. The main problem is the fact that the number of these bodies must eventually be considerably reduced (i.e. from swarms of small bodies, eventually a few large planets must be produced). This reduction can only take place by collisions. Although little is known about the "sticking" probability of rocks *in vacuo*, it seems that at the typical speeds expected for bodies in Keplerian orbits around the sun ($\simeq 10$ km/sec), collision of rocks is not very likely to lead to coalescence.

The model of Safronov (1969, 1972) starts with a quiet nebula in which dust particles rotate and begin to sink slowly (under gravity) towards the equatorial plane. The assumption is made that all the particles such a dust particle meets on the way down adhere to it, and, upon arriving at mid-plane (after $\simeq 10^3$–10^4 yr) it has reached a size of $\simeq 1$ cm. The critical assumption is the so called "cold welding" of matter which represents an upper limit to the efficiency of the process. When the density of the dust layer exceeds a certain value, condensation of matter would be induced by gravitational instabilities.

In those regions close to the sun, gas motions would have been too fast for gravitational instabilities to take place and growth of bodies could only occur through aggregation during collisions. Safronov argued that the relative velocities of the bodies would be determined by their gravitational perturbations at encounters and should be less than 1 m/sec so long as the objects were less than 5 km in diameter. This velocity is 10^{-4} less than the Keplerian velocity. Alfvén and Arrhenius (1970a, b, 1973) solved the problem of reducing the relative velocities of colliding bodies so as to facilitate coalescence by the concept of "jet" streams. They argued that collisions between particles in Keplerian orbits do not lead to a spreading but to an equalization of the orbits of the particles when the collisional frequency is smaller than the orbital frequency. This state, in any theory of this type, either prevails originally or develops by accretion. It results in a focusing of particles into jet streams with increasingly similar orbital elements and velocities of the individual particles. The partially inelastic collisions will lead to the growth of embryo bodies of which some eventually reach the size where gravitational accretion becomes increasingly important. Alfvén and Arrhenius believe that the maximum possible velocity for accretion is 0·5 km/sec. It must be pointed out that the jet stream hypothesis of Alfvén and Arrhenius has not been universally accepted (see e.g. Reeves (1972), p. 86–87, for a discussion on this point).

Cameron (1972) has also recognized that during the lifetime of the nebula, collision probabilities are too low for accretion to take place in any reasonable

c

time. He suggested however that during the collapse phase from the interstellar cloud to the nebula the situation was much more favourable since violent gas motions (in the form of turbulent eddies) would have considerably increased the collision probabilities of gas and grains. Cameron estimated typical grain velocities to be $\simeq 10$ m/sec. At the end of the collapse the nebula is formed and turbulence rapidly decays out embedding a set of bodies up to 20 cm in radius.

In its initial, non-gravitational stage, the accretionary process may be regarded as a series of collisions between grains in intersecting orbits under conditions which result in a net transfer of mass to some of the grains. An important parameter for accretion theories is the range of impact velocities over which accretion occurs. Kerridge and Vedder (1972) investigated experimentally this velocity range by observing the impact at different velocities of particles similar to material thought to be present in the early solar system. The projectile material used was kaolinite (a hydrated aluminum silicate) in the form of thin micrometer-sized laminated flakes with a density of 2.63 g/cm³. Such material structurally resembles the major mineral component of the primitive type I carbonaceous chondrites and may well have been widespread in the early solar system. For target material, thick ($\simeq 5$ mm) sections of clinochrysotile (a magnesium silicate) with a density of 2.50 g/cm³ were used. They found that in any velocity range, impacts in which a significant transfer of projectile material occurred were outnumbered by those in which little or no transfer took place. Moreover such impacts, at least those above 2 km/sec, were accompanied by the formation of craters with dimensions comparable to or greater than those of the projectile. Thus no accretion could take place for impacts in the velocity range studied (1.5–9.5 km/sec). By contrast, Neukum (1968) showed that for micrometer-sized iron particles impacted into a variety of metallic targets, almost the entire projectile mass is transferred to the target in the velocity range 0.5–13 km/sec. Below 0.5 km/sec iron particles rebounded from the metallic targets. There is thus a striking difference between the accretionary behaviour of silicates and that of metals—over at least part of the velocity range in which silicate accretion was efficient, metal particles would have rebounded without accreting.

Simple accretionary theory cannot, therefore, be applied to silicate particles in markedly dissimilar orbits but must be restricted to situations characterized by generally low inter-particle velocities. Whether or not impacts at lower velocities can actually lead to accretion of material is not known. No obvious sticking mechanism exists for bonding normal silicates together during low velocity impact, although a number of mechanisms have been proposed based on assumed properties of primordial dust grains. Possibly the sticking process was aided by the presence of volatile coatings on the grains. Alternatively, adhesion may have been promoted by electrostatic charge asymmetries produced in the surface regions of circumsolar grains by charged particle irradiation, as proposed by Arrhenius *et al.* (1972). A third possible mechanism, suggested by Maurette and Bibring (1972), is the release of stored energy from the irradiation-damaged amorphous rims of grains exposed to the solar wind.

Kerridge and Vedder (1972) suggested that metal–silicate fractionation in the solar system may have been affected by differences in the accretionary behaviour of metal and silicate particles. In particular at the low velocities which must have prevailed during most of the accretionary process, silicates would have preferentially accreted. Similarly, if some regions of the solar system were characterized by higher inter-particle velocities, bodies enriched in metal may have been produced. In this respect it may be significant that among the terrestrial planets there is a rough tendency for density, and therefore assumed metal content, to increase with orbital velocity.

A problem common to all models of cold origin is the means by which the Earth could have heated up sufficiently to lead to a (predominantly iron) molten outer core at least 2700 m yr ago. This problem will be discussed in the next section.

2.3. Heat Sources for an Earth Accreting Cold

(i) Long-Lived Radioactive Isotopes

The radioactive isotopes which contribute significantly to the present heat-production within the Earth are U^{238}, U^{235}, Th^{232}, and K^{40}, all of which have half lives comparable to the age of the Earth. The temperature increase due to the radioactive decay of these long-lived isotopes is thus small, of the order of 150°C after 100 m yr (MacDonald, 1959). During the first 1000 m yr, the temperature increase would only be about 700°C (assuming no heat escape) while the total heat produced over the life time of the Earth (~ 4500 m yr), if trapped within the Earth, would raise the temperature by ~ 1800°C. Thus long lived radioactive isotopes can account for part of the initial heating of the Earth, but other sources are necessary as well.

(ii) Short-Lived Radioactive Isotopes

Short-lived radioactive isotopes could have contributed to the initial heat of the Earth if the time between the formation of the elements and the aggregation of the Earth was short compared with the half lives of the isotopes. The most important short-lived isotopes are U^{236}, Sm^{146}, Pu^{244}, and Cm^{247}, all of which have half lives sufficiently long to have heated up the Earth for some tens of million years after the initial formation. These four isotopes would, in fact, have contributed about 20 times the heat produced by K^{40} during this period. The decay of three shorter-lived radionuclides Al^{26}, Cl^{36} and Fe^{60} would have significantly heated up accreting planetary bodies for a period of about 5–15 m yr after the termination of nucleosynthesis in the primitive solar system (Fish *et al.*, 1960). Of these, Al^{26} is the most important. It decays to Mg^{26} with a half life of 0·74 m yr and would remain as a significant source of heat for about 10 m yr. If the Earth accreted within 20 m yr of the termination of nucleosynthesis, the heat released through the decay of Al^{26} could be the main cause of its high internal temperature. If, on the other hand, the time of accretion was of the order of 100 m yr, the decay of Al^{26} would have had a negligible effect on the Earth's thermal history. In this regard, no trace of any anomaly in the Mg^{26}/Mg^{24} ratio had, until recently, been found in meteorites or in lunar and terrestrial samples (Schramm *et al.*, 1970). Gray and Compston (1974), however, have found an anomalous value of the Mg^{26}/Mg^{24} ratio in a melilite-bearing chondrule in the Allende meteorite. They suggest that the anomaly results entirely from excess Mg^{26}, implying an origin by the radio-active decay of Al^{26} formed during intense proton irradiation in the early stages of the solar system. Lee and Papanastassiou (1974) have also analysed samples of the Allende meteorite and found Mg^{26} anomalies—in one of the samples, however, the anomaly is negative which appears to rule out *in situ* decay of Al^{26} as the cause of the anomalies. There seems to be no doubt now of the presence in meteorites of the products of nucleosynthesis which have not been thoroughly mixed, but the exact process involved is not at all clear.

Some information on the time of nucleosynthesis has come from a study of xenon isotopes. I^{129} decays through β-emission to Xe^{129} with a half life of 16·4 m yr. This half life is so short that no I^{129} now exists in nature. However this isotope should be produced at the time of formation of the other elements. It can be expected that the condensation of planetary material would incorporate any of the iodine present, but would be unlikely to incorporate much xenon. Xe^{129} formed within the condensed materials would be trapped and might be observed today—very small amounts of excess Xe^{129} have in fact been found in a number of meteorites. Absolute formation intervals calculated by the Xe^{129}–I^{129} method generally give values between 40 and 300 m yr depending on the nucleosynthesis model assumed. Variations in relative formation intervals are much less, but are difficult to assess because of different methods of data treatment between different laboratories. Podosek (1970) has subjected all of the Berkeley data to a common method of analysis and found 15 different meteorites possessing formation intervals within a period of 15 m yr.

Other noble-gas components in meteorites that arise from an extinct radionuclide and have the potential for the determination of formation intervals are the heavy xenon isotopes $Xe^{131-136}$ which arise from the fission of Pu^{244}. These fission Xe components are much more difficult to identify than the excess Xe^{129} and were not discovered until precise isotopic data were obtained on the calcium-rich achondrites, a class of meteorites high in uranium (supposedly chemically similar to plutonium) and very low in trapped Xe (Rowe and Kuroda, 1965).

Hoffman et al. (1971) claim to have discovered Pu^{244} in nature—in an old Precambrian rare-earth mineral called bastnaesite from the Mountain Pass deposit, California. Pu^{244} has a half life of about 82 m yr so that the solar system was formed about 60 half lives ago. The difficulty of finding natural Pu^{244} is that its average concentration should be only $(\frac{1}{2})^{60}$ of its original value and it could thus only hope to be found in some mineral in which it was enriched several orders of magnitude. The discovery, if substantiated, would have profound cosmological significance. Considered in conjunction with the primeval meteoritic abundance of I^{129} it can be concluded that about 180 m yr elapsed from the time when a portion of interstellar gas collapsed into the gas cloud that was to form the solar system and the time when the newly formed meteorites were cool enough to retain noble Xe gas. Hoffman et al. point out that the present abundance of Pu^{244} due to the influx of Pu^{244} nuclei in cosmic rays may be comparable with, or even greater than, that surviving from primordial Earth material. In this respect, Fleischer and Naeser (1972) have found that the Mountain Pass bastnaesite has an apparent Cretaceous fission track age and thus does not reveal any anomalous fission tracts due to Pu^{244}. Yet a third possibility for the existence of Pu^{244} in nature has been put forward by Sakamoto (1974). He finds difficulties with both the possibilities of the survival of primeval Pu^{244} or of an influx of a heavy cosmic ray component. Instead he suggests it is the result of influx of a cosmic dust component from supernova remnants. Although it is speculative whether any part of the cosmic dust falling on the Earth originated outside the solar system, preliminary calculations by Sakamoto indicate that the Pu^{244} may have been carried by such dust. Further work is clearly necessary on this very important matter.

Rao and Gopalan (1973) have also considered the fission Xe components in different types of meteorites and in particular have investigated whether some actinide isotopes could produce the excess fission Xe found in primitive chondrites. They concluded that the I–Xe and Cm–Xe systematics relate to different events in the early history of the solar system. The former refers to the time when I–Xe clocks are re-set during metamorphism of the ordinary chondrites while the latter refers to the time of primary condensation of the refractory materials from the cooling solar nebula (Hoyle and Wickramasinghe, 1968). The occurrence of the short-lived nuclide Cm^{248} ($\alpha T_{1/2} = 0.4$ m yr) in primitive meteorites has important consequences for the time scales of the earliest events in the evolution of the solar system. As the solar nebula cools from about 2000 to 1400°K, i.e. to the temperature at which iron and magnesium silicates condense, the high temperature minerals (or phases) form as refractory condensates (Fireman *et al.*, 1970; Podosek and Lewis, 1972). Curium, being a refractory element similar to U, Pu and the rare-earths, is most likely to be incorporated into these minerals which would preserve the spontaneous fission record of Cm^{248} if they cooled fast to the Xe retention temperature. Hence this actinide isotope could fix the time scale for a fast cooling stage, i.e. if the time interval for condensation was about 2 m yr, the Cm^{248} record would be preserved in these early condensations, but if it lasted for tens of m yr, Cm^{248} would have completely decayed before it could be incorporated into these minerals.

(iii) Adiabatic Compression

The temperature of the material within the aggregating Earth would also increase because of adiabatic compression. Although data (particularly on the variation with pressure of the coefficient of thermal expansion) are rather uncertain, a rise in temperature of several hundred degrees from this source seems possible.

(iv) Potential Energy due to the Mutual Gravitational Attraction of the Particles of the Dust Cloud

The kinetic energy of the aggregating particles is either converted into internal energy or radiated away. It is extremely difficult to estimate the contribution from this source because of our lack of knowledge of the physical processes of accretion. The result depends quite critically on the temperature attained at the surface of the aggregating Earth and on the transparency of the surrounding atmosphere to radiation.

Comparatively low surface temperatures (of the order of a few hundred degrees) have been suggested, mainly because the atmosphere of the primitive Earth was assumed transparent so that the thermal energy of the impinging particles was immediately re-radiated into space. Ringwood (1960) has argued, however, that during these early years, the primitive Earth would have a large reducing atmosphere. In the presence of these reducing agents (chiefly carbon and methane), the accreting material would be reduced to metallic alloys—principally of iron, nickel and silicon. The outer regions of the Earth would thus be metal rich and dense (referred to zero pressure) compared with the interior. Such a state is gravitationally unstable, and convective overturn would follow leading to a sinking of the metal rich outer regions to the centre. This would release further heat due to the energy of gravitational rearrangement. Ringwood believes the whole process is likely to be catastrophic, since the overturn would be accelerated as the initial temperature rose.

The relative importance of the gravitational potential energy of the dust cloud as a means of heating the proto-Earth depends critically on the duration of the accretion process—rapid accretion is necessary to produce high temperatures and melt the outer core. The rate of growth of mass of the embryo Earth would increase in the beginning of the accretion process (because of the increase in the capture cross-section of the embryo), pass through a maximum and then decrease (because of the exhaustion of accretable material).

The law governing the flux, $\dot{m}$, of falling matter determines the ultimate thermal profile since losses by radiation depend upon how fast the recent fall is covered by new layers thus insulating the former from radiative loss. The energy balance is given by the equation

$$\frac{GM(t)\dot{m}}{r(t)} = \epsilon_m \sigma \left[T^4(t) - T_0^4\right] + c_p \left[T(t) - T_0\right]\dot{m} \tag{2.1}$$

for infalling matter having the velocity of escape at impact, where G is the universal constant of gravitation, $M(t)$ the mass of the Earth, $\dot{m}$ the mass flux, $r(t)$ the instantaneous radius of the Earth, ρ the mass density, ϵ_m the emissivity, σ the Stefan–Boltzmann constant, $T(t)$ the instantaneous temperature of the surface of the proto-Earth at time t, T_0 the background temperature of free space of the circumsolar nebula into which radiative losses take place, and c_p the specific heat at constant pressure. In Eqn. (2.1) the term on the left-hand side represents the gravitational energy input, the first term on the right-hand side the radiative loss rate, and the second term the rate of heat takeup by the infalling matter in coming to the temperature, $T(t)$. Heat conduction to the interior from the hot surface regions is neglected since it is vanishingly small for time scales of the order of 10^3–10^6 yr. Heating due to adiabatic compression is also ignored.

The mass flux, $\dot{m}$ (t), is the dominant parameter in calculating the thermal profile resulting purely from accretion, although T_0 has an important effect under certain conditions. In general $\dot{m} = \dot{m}(t)$ but, as will be shown later, whether $\dot{m} = \dot{m}(t)$ or is constant has little bearing upon the main conclusions. It is easy to generalize a set of laws for $\dot{m}$ which include gravity and slower rates of accretion as well. It can be shown that the exact form of the law has little effect upon the time limit for sufficient energy to be stored to melt surface or subsurface regions. Write

$$\dot{m} = c_n/r^n \tag{2.2}$$

where $r = r(t)$ and c is a constant with index n. Since $\dot{M} = 4\pi\rho r^2 \dot{r}$,

$$\dot{m} = \dot{M}/4\pi r^2 = \rho\dot{r}. \tag{2.3}$$

Hence from Eqns. (2.2) and (2.3),

$$\rho \, r^n \, \dot{r} = c_n.$$ (2.4)

This integrates to

$$r^{n+1} - r_0^{n+1} = \frac{(n+1) \, c_n}{\rho} (t - t_0)$$ (2.5)

except for the case $n = -1$. In this case

$$\ln\left(\frac{r}{r_0}\right) = \frac{c_{-1}}{\rho} (t - t_0).$$ (2.6)

For $n = 0$, $\dot{m}$ is constant and the radius grows linearly with time while for $n \geqslant 1$ the growth is slowed. None of these assumptions is especially plausible physically, as the accretion rate should depend upon the gravitational force ($n = -1$) or potential ($n = -2$), assuming that the escape velocity is not exceeded and an infinite source reservoir is available.

For the case $n = -2$, Eqn. (2.1) becomes

$$\tfrac{4}{3}\pi\rho G c_{-2} \, r^4 = \epsilon_m \sigma \, (T^4 - T_0^4) + c_p \, c_{-2} \, r^2 \, (T - T_0)$$ (2.7)

where the constant $c_{-2} = \dot{m}_f / R_m^2$ where $\dot{m}_f$ is the mass flux at the termination of growth, $r = R_m$. From Eqn. (2.5), it then follows that

$$\frac{1}{r_0} - \frac{1}{r} = \frac{\dot{m}_f}{\rho R_m^2} (t - t_0)$$ (2.8)

and the time taken for the Earth to grow from radius, r_0, to final radius, R_m is given by

$$t = \frac{\rho R_m^2}{\dot{m}_f} \left[\frac{1}{r_0} - \frac{1}{r} \right] = \frac{\rho R_m^2}{\dot{m}_f} \frac{\Delta r}{r r_0}$$ (2.9)

where $\Delta r = r - r_0$. When $r = R_m$,

$$t = \frac{\rho R_m}{\dot{m}_f} \left(\frac{\Delta r}{r_0} \right) = \frac{\rho R_m}{\dot{m}_f} \left(\frac{R_m}{r_0} - 1 \right).$$ (2.10)

For $R_m/r_0 = 2$, $t = \rho R_m / \dot{m}_f$ which is the same as for the case $\dot{m} = c_0$. Thus the time taken for the Earth to grow from half radius to its full value under gravitational accumulation is the same as for $\dot{m} = c_0$. Although the radiative losses depend upon the accretion time, there is not a great difference in the final temperature for these two cases. The effect of heat being admitted to the incoming matter tends to diminish the differences between the different mass laws for $\dot{m}$. Without this term in Eqn. (2.1), the temperature would vary as r for gravitational accretion and be less for cases where $n > -2$. If all the gravitational energy of accumulation were trapped in the Earth, the first term on the right-hand side of equation (2.1) vanishes and the resulting increase in temperature ΔT can be estimated from the equation

$$\frac{GM}{R_m} = c_p \, \Delta T \tag{2.11}$$

where R_m is the final radius. This gives a temperature of the order of 50,000°K. This is the maximum thermal effect of the accretion process— such a hot origin of the Earth would lead to core formation and large scale differentiation upon accretion (Ringwood, 1966).

The minimum thermal effect of the accretion process may be obtained by assuming that the energy released by inelastic collisions of matter falling in from the primordial gas–dust cloud is partially compensated by energy re-radiated to space. The equilibrium condition is then given by Eqn. (2.1) with the last term omitted. If we assume black-body radiation, $\epsilon_m = 1$ and the resulting temperatures are minimal. The unknown quantity is $\dot{r}$, the rate of accretion. Hanks and Anderson (1969) have considered a number of models. The simplest models assumed a constant accretion rate: more plausible models were obtained with accretion rates $\dot{r} = A \sin \alpha t$ and $\dot{r} = C t^2 \sin \gamma t$, where

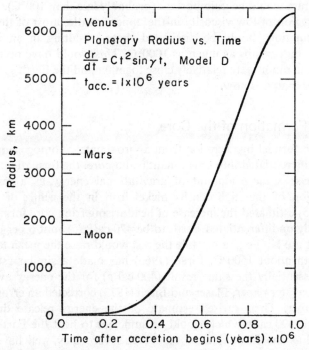

Fig. 2.1. Radial growth of the Earth. (After Hanks and Anderson, 1969.)

$\dot{r}$ is zero at $r = 0$ and $r = R_m$. The constants A, α, C and γ are determined by the final radius R_m and t_{acc}, the total time of accretion.

Figure 2.1 shows the radial growth as a function of time for this last model, when $t_{acc} = 10^6$ yr. It can be seen that it takes 0.3×10^6 yr for the Earth to reach asteroidal size—the rapid acceleration of the accretion process between 0.4–0.9×10^6 yr is responsible for the major part of the temperature rise. The accretion rate begins to decelerate significantly only in the last 200 km. Safronov (see, e.g. 1969) and Urey (1962) prefer an accretion time of the order of 10^8 yr, rather than the very much shorter times advocated by Ringwood (1959, 1966) Hanks and Anderson (1969) and Turekian and Clark (1969).

(v) Dissipation of the Earth's Rotational Energy

As the Earth slows down through tidal interaction with the moon (and to a lesser extent with the sun), part of its rotational energy is dissipated by tides in the oceans and part in the interior of the Earth by Earth tides. It is difficult to estimate how the energy loss is divided between ocean and Earth tides— probably most of it is dissipated in the shallow seas. If the heat generated by tides were evenly distributed over the whole volume of the Earth, the increase in temperature would be unimportant (probably less than 100°C). However if there is a region of low viscosity in the upper mantle, most of the energy of tidal deformation should be dissipated there, resulting in an increase of temperatue by perhaps as much as 1000°C. This would have occurred when the Earth and moon were relatively close together and the Earth's rotation was much faster than it is now.

(vi) The Formation of the Core

If the Earth formed by accretion from approximately homogeneous material and later differentiated into crust, mantle and core, the formation of the core would release a large amount of gravitational energy as a result of the concentration of the high density nickel–iron in the centre of the Earth. Tozer (1965) estimated the increase of heat arising from core formation from an originally undifferentiated Earth to be 470 cal/g. About 6 per cent of this would melt the Ni–Fe phase, while the rest would raise the mean temperature of the Earth about 1500°C. Birch (1965) has made a similar estimate and obtained essentially the same result (400 cal/g) for the energy available for heating. In a later paper, Flaser and Birch (1973) corrected an error in Birch's original paper. Their revised estimate of the energy release due to core formation is 590 cal/g which would be sufficient to heat the Earth by about 2000°C. Thus the formation of the Earth's core may well have played a dominant role in the thermal history of the Earth.

(vii) Impacts of Falling Bodies

Safronov (1968, 1972) is of the opinion that one of the main sources of heat for an Earth accreting cold was the impacts of falling bodies. Estimates of the initial temperature from this source depend largely on the body sizes assumed. Calculations based on the assumption that these bodies were small, give a low initial temperature—from $300°K$ near the surface to $800–900°K$ at the centre. The impact energy of small bodies and particles would be liberated near the surface of the growing Earth, practically all of it being radiated back into space. Safronov believes however that relatively large bodies were involved in the formation of the planets, the largest bodies falling on the Earth having diameters of several hundred km. The larger the incident body, the greater the depth at which its impact energy would be released and hence the greater the fraction of this energy trapped inside the Earth, unable to escape into space. Moreover, larger bodies would produce deeper craters, inducing more intensive mixing of the material on impact. Heat transfer by mixing of material during the impact of large bodies is far more efficient than heat transfer by ordinary thermal conduction. Numerical estimates show that the maximum initial temperature of the Earth occurred in the region of the upper mantle (at a depth ~ 500 km) and probably exceeded $1500°K$.

The largest bodies striking the Earth would also cause primary inhomogeneities in the mantle. Apart from inhomogeneities arising from differences in chemical composition of the impacting bodies, there would also be temperature inhomogeneities. Regions of impact of the largest bodies with sizes up to 1000 km would be additionally heated some hundreds of degrees —in such large areas the temperature would not be equalized even after a 1000 m yr. With the additional heat due to radioactive decay, the central zones of these regions would be the first to reach melting temperatures. When about half of the material had melted, local gravitational differentiation would take place—heavy materials would settle down to the lower part of the zone, and light materials rise to the upper part. During such global differentiation large amounts of thermal energy would be released. As a result a greatly heated vertical column of lower viscosity would be formed along which masses of heavier materials would sink downwards to form the core. Safronov also suggested that it was above such regions excessively heated by impacts that the formation of the continents took place.

2.4. Time of Core Formation

Views of the time required to form the Earth's core have changed considerably over the past two decades. Runcorn (1962a, b) suggested that the core has continued to grow throughout geologic time and developed a theory of continental drift based on its growth. Vening Meinesz (1952) had argued that the positions of the continents today could result from a large scale, regular pattern of convective motions in the mantle, continental material tending to congregate at places where the currents are descending. Using the results of Chandrasekhar's studies (1961) on the convection of a fluid in a spherical shell, Runcorn showed that the present radius of the core is only just greater than the value at which the fourth harmonic convection pattern in the mantle should become less readily excited than the fifth. During a transition from one harmonic to another the continents will be under much stress as the convection pattern changes to a new form. The relatively recent time (geologically speaking) at which continental drift took place is thus understandable. Runcorn also predicted other epochs of large continental displacements when the core radius reached successively the values at which the first degree convection pattern in the mantle, initially present when the core was only beginning to form, gave place to the second and the second to the third, etc. These other epochs of continental drift occurred early in the early Precambrian.

In addition to continental displacements one would expect a series of orogenies during such transition periods. Radioactive age determinations have mainly been obtained from igneous rocks which come from depth or metamorphic rocks formed in the deeper parts of the crust as the result of orogenic forces (Gastil, 1960). It is interesting that the ages obtained for such rocks group in broad peaks around the dates 2600, 1800, 1000, and 300 m yr ago. The most recent peak covers much of the geologic record since the middle Palaeozoic and this is associated with continental drift and the transition from a fourth to a fifth degree convection pattern in the mantle. Runcorn suggested that it is natural to identify the 1000 m yr peak with the transition from the third to the fourth degree convection pattern, the 1800 m yr peak with the transition from the second to the third and the 2600 m yr peak with the transition from the first to the second. In this way the dates at which the core radius reached these critical values can be determined—see Fig. 2.2 which also shows that, on this theory, the core only started its growth a little over 3000 m yr ago.

Elsasser (1963) has considered in some detail the model of an Earth accreted cold with the material uniformly distributed. The main feature of such an Earth model is that the melting point curve of the silicates rises much more steeply with depth than the actual temperature. This implies that the viscosity of the silicates should increase appreciably with increasing depth. As the original Earth is heated by radioactivity the outer layers are then the first to become soft enough to permit iron to sink towards the centre—farther down the fall of iron is slowed by increased viscosity. It then forms a coherent layer which, however, is gravitationally unstable and results in the formation of quite large drops. The latter fall rapidly to the centre giving rise to a proto-core. Elsasser estimated that the formation of a proto-core slightly smaller than the present core probably took no more than several hundred m yr.

The fall of iron is controlled by the elasto-viscous properties of the silicate matrix. Below the melting point, the viscosity will increase exponentially with decreasing temperature. Thus instead of a uniform rain of iron towards the centre we have a far more complicated process in which the fall of iron is largely controlled by the variation in temperature at any given depth. As iron

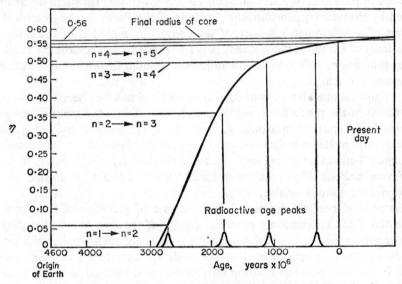

Fig. 2.2. Growth of the Earth's core compared with radioactive age determination peaks. (After Runcorn, 1962b.)

fell through the almost fluid silicate material the general flow pattern of the latter would at first be nearly streamline, but, since the viscosity of the mantle increases with depth, the inner portion would not have time to flow but would be pushed outwards by the falling drop of iron. Elsasser suggested that the composition of the mantle might thus become asymmetrical, the material displaced to the antipodal point containing iron, while that above the falling drop lost most of its iron.

Tozer (1965) has reconsidered the question of the kinetics of core formation and concluded that the simple theory of falling iron masses in silicate material is untenable. He suggested as an alternative a mechanism based on the flow of iron along channels in the silicate phase—the acceptability of this theory however depends quite critically on whether iron is able to flow over distances of the order of a km under such conditions. Tozer concluded that in any case core formation is proceeding today much more slowly than in the past and that it was virtually complete very early in the Earth's history.

Tolland (1973) has also considered core formation following accretion from an initially homogeneous mixture (82 per cent by volume of silicate and 18 per cent of iron plus sulphide). His model differs from other models (such as that of Elsasser, 1963) in that he considers the role of convection during core formation, taking into account the interaction between infall and convection processes. Tolland used Tozer's (1972) model of the thermal history of the Earth (see Section 3.6). Tozer showed that for small departures from hydrostatic equilibrium, the rheology of the Earth can be described in terms of a Newtonian viscosity. If the convecting region is of the order of hundreds of km thick, the viscosity is $\sim 10^{20}$ poise for any reasonable value of heat generation. This answers the objection to infall theories that the mantle viscosity is too high.

Tolland assumed the present core to be an Fe–S mix (see Section 5.5)—the first melt in the proto-Earth would then be well below the melting point of iron. Rapid convection would prevent the Earth from ever having been extensively melted—a fact which Urey (1952) deduced on geochemical grounds. Tolland estimated that, for a mantle viscosity of 10^{20} poise, to form the core in about 10^8 yr requires an initial "seed" of about 1 m in size, which does not seem unreasonable.

As pointed out in Section 2.3, a rapid rate of accretion of the Earth is essential if the gravitational potential energy of the dust cloud is to play a major role in the heating of the embryo Earth. There are a number of lines of evidence which indicate that the time scale for accretion was quite short.

It is generally believed that the Earth's magnetic field arises from dynamo action in the fluid, electrically conducting outer core (see Section 4.2). Remanent magnetization has been found in rocks as old as 2700 m yr so that the outer core must have been molten at least that long ago. Hanks and Anderson (1969) have carried out an investigation of the thermal history of the Earth with this additional constraint and come to the conclusion that the Earth must have accreted in a period less than 0·5 m yr. There would not have been time for the decay of long lived radioactive isotopes to have had sufficient heating effect and the authors discount the importance of the decay of short-lived radionuclides. A short accretion time is thus essential so that most of the gravitational potential energy of the dust cloud is trapped and not lost to space. The accretion time could be lengthened somewhat since the core is not pure iron. The addition of silicon or sulphur (see Section 5.5) would lower the melting point of iron—Hall and Murthy (1972) have estimated that at core pressures the melting temperature of the Fe–FeS eutectic is about 1600°C lower than that of pure iron.

Wetherill (1972) has pointed out that measurements of the concentration of Rb and Sr and the Sr isotopic composition of lunar samples indicate that a major fractionation of Rb relative to Sr took place on the moon 4600 m yr ago. Similar studies of the highly differentiated lunar breccias provide additional evidence for a very early lunar differentiation. Wetherill concluded that the most likely heat source for this initial differentiation is the gravitational energy of lunar accretion and that, for a sufficiently high temperature to be reached during the accretion period, the time scale for accretion must be of the order of 1000 yr. By analogy with the moon Wetherill suggested that the entire Earth was initially melted and fractionated, and that the age of the oldest rocks ($\simeq 3500$ m yr) indicates that it was not until then that the Earth cooled sufficiently to enable the formation of extensive areas of stable crust.

Oversby and Ringwood (1971) have also produced evidence for an early formation of the Earth's core. Ringwood had argued in 1960 that iron descending during core formation would take with it substantial amounts of lead but not uranium, i.e. core formation would change the Pb/U ratio in the upper crust and mantle. He thus concluded that the 4550 m yr terrestrial event recorded by Pb/U geochronology referred to the time of core formation which must have taken place very soon after the formation of the Earth. The main argument for Ringwood's conclusion is that during core formation the Pb/U ratio of the metal phase would be higher than that in silicates; the opposite point of view was taken by Patterson and Tatsumoto (1964). Oversby and Ringwood (1971) thus carried out experimental measurements of the distribution coefficient of lead between relevant metal and silicate systems to settle this question.

Since the core is believed to contain some 10–30 per cent of one or more light elements in addition to iron (see Section 5.5), they examined a number of core models with different synthetic metal phase compositions ($Fe_{83}S_{15}C_2$, $Fe_{89}Si_{11}$, $Fe_{30}S_{30}$—weight per cent). The silicate phase used in the experiments consisted of simplified synthetic basaltic compositions, doped with 1000 ppm of lead (as oxide) in some runs and 100 ppm in others. Finely powdered samples of metal phase were mixed with silicate in the ratio 1 part (by weight) of metal to 2 parts silicate. The mixture was heated to a temperature at which both metal and silicate phases were known to be molten (the experiments were carried out at high pressures under closed system conditions to prevent volatilization of lead). Their results showed that lead ends up preferentially in the metal phase and values of the distribution coefficient $K^{m/s}$ (i.e. the ratio of Pb concentration in the metal to that in the silicate phase) were obtained for each run.

A relationship exists between $K^{m/s}$ and the time interval ΔT_c between accretion of the Earth and segregation of the core. The exact relationship

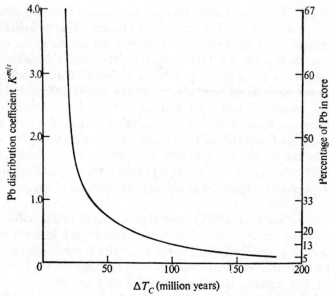

Fig. 2.3. Relationship of the time delay for core formation after accretion of the Earth, ΔT_c, with the distribution coefficient for lead between metal and silicate phases, $K^{m/s}$. Time of accretion is taken as 4550 m yr. Isotopic composition of upper mantle lead is taken as that from tholeiites dredged from the floor of the Pacific Ocean. (After Oversby and Ringwood, 1971.)

depends on certain "boundary" conditions—Fig. 2.3 shows one of the curves produced by Oversby and Ringwood. The distribution coefficient for a model with a metal phase composition $Fe_{83}S_{15}C_2$ is 2·5 and Fig. 2.3 shows that core segregation must have occurred within about 20 m yr after accretion. For the $Fe_{70}S_{30}$ and $Fe_{89}Si_{11}$ compositions, $K^{m/s}$ values are larger—however, because the ΔT_c curve is very steep above $K^{m/s} = 1·5$, the ΔT_c intervals are only slightly shorter. Oversby and Ringwood thus concluded that the Earth's core formed either during accretion or very soon after. Reasonable variations in the boundary conditions do not affect this basic result.

These conclusions strongly favour models in which much of the accretion of the Earth occurred rapidly under sustained high temperature conditions, rather than those in which the Earth accreted in a cool, unmelted state, followed by subsequent internal heating and core formation at a much later stage.

2.5. Inhomogeneous Models of the Earth

Orowan (1969) was one of the first to suggest that the Earth may have accreted inhomogeneously. He pointed out that iron is plastic ductile, even at low temperatures, provided that it does not contain far more carbon than is found in meteorites. As a result metallic particles would be expected to stick together when they collide because they can absorb kinetic energy by plastic deformation. They can therefore combine by cold or hot welding. Silicates, on the other hand, are brittle and break up on collision except within a narrow temperature range near their melting point. The accretion of the planets may thus have started with metallic particles. Once sufficiently large a body could easily collect non-metallic particles by embedding them in ductile metal, and later by gravitational attraction. Orowan thus suggested that planets may arise cold in this way with a metal core already partially differentiated. The problem of heat sources to provide subsequent melting of the core still remains however. Harris and Tozer (1967) pointed out that in the preplanetary dust cloud, particle adhesion and aggregation would have been most effective for magnetic grains. They thus suggested that the magnetostatic attraction of ferro-magnetic dust particles could lead to enhanced capture or "sticking" cross-sections. They showed that with a Curie point for iron or an iron–nickel alloy of more than 700°C and an average time interval between dust particle collisions (based on the optical cross-section) of the order of one year, there would have been time for a very large number of such collisions before oxidation of the iron occurred to any significant degree.

Eucken (1944), Anders (1968), Turekian and Clark (1969) and Clark *et al.* (1972) have also proposed a model Earth in which the Earth accreted in-homogeneously—not a cold accretion model like that of Orowan (1969), but as a result of condensations from the solar nebula. The observed decrease in density of solid bodies with distance from the sun may then be causally related to the radial temperature gradient present in the solar nebula during the condensation of the component materials. The sequence of chemical compositions of solar system bodies may be pictured as paralleling the sequence of condensation reactions undergone by a solar composition gas during cooling from > 2000°K to perhaps 30°K. More specifically, it is possible in principle to calculate the equilibrium chemical composition of solar material as a general function of pressure and temperature, and specify the exact sequence of reactions and the composition of the condensate along any pressure–temperature profile. Larimer (1967) calculated that the order of condensation would be Fe and Ni, magnesium and iron silicates, alkali silicates, metals such as Ag, Ga, Cu, iron sulphide and finally metals such as Hg, Tl, Pb, In and Bi. Such an order of condensation is that grossly inferred in the Earth and usually attributed to differentiation. Turekian and Clark (1969) thus suggested that the Earth's core formed by accumulation of the con-densed Fe–Ni in the vicinity of its orbit which, as in Orowan's model, then became the nucleus upon which the silicate mantle was deposited. However Blander and Katz (1967) and Blander and Abdel-Gawad (1969) maintain that, at the pressures expected to have existed in the solar nebula, the silicates would have condensed before iron. Larimer has now revised his earlier estimates and also believes that iron condensed slightly later than the silicates (Anders, 1968; Larimer and Anders, 1970).

Lewis (1972a, b) has developed a model for the origin of the solid bodies in the solar system beginning with a solar nebula in which there is a steep radial temperature gradient. He assumed that the bulk composition of condensates in the nebula is determined by chemical equilibrium between the condensates and gases in a system of solar composition. He then calculated the bulk density of condensate as a function of temperature (over the range

Table 2.1. Condensation sequence for solar material. (After Lewis 1973).

Reaction	Temperature (at 10^{-3} bar) (°K)	Cumulative bulk density of condensate (g/cm³)
Refractory oxide condensation	1720	~3·5
Metallic iron condensation	1460	~7·0
MgSiO₃ condensation	1420	4·40
Alkali aluminosilcate formation	1250	4·40
FeS formation	680	4·46
Tremolite formation	540*	~4·3*
End of Fe oxidation to FeO	490	3·85
Talc/serpentine formation	400	3·2

* Tremolite formations occur while the density of the condensate is already falling rapidly due to Fe oxidation. A small uncertainty in the free energy of formation of tremolite thus corresponds to a large uncertainty in the bulk condensate density at the temperature of tremolite formation.

0–2200°K) and pressures from 10^{-7} to 10^{+1} atm. He found that the graph of density of condensate versus temperature was quite insensitive to total pressure over a wide range of pressures, the sequence of reactions and compositions of solid phases formed at equilibrium being essentially pressure-invariant. The condensation and reaction sequence for inner solar system material of the most abundant elements is given in Table 2.1.

Lewis obtained a specific density versus temperature curve (Fig. 2.4) which shows large density decreases due to oxidation of iron, hydration of silicates, condensation of water ice, and formation of solid gas hydrates. A notable exception to the trend of decreasing density with falling temperature is the formation of FeS, the mineral troilite, at 680°K. Retention of the volatile element sulphur leads to a density increase because of the high atomic weight of sulphur. Lewis' model also permits specific correlation of the total volatile content of the bodies with their bulk densities. Thus density, oxidation state, and volatile content are all causally interrelated.

The observed density trends of the terrestrial planets thus need not be the result of special fractionation processes, but a consequence of physical and chemical restraints on the structure of the solar nebula, particularly the variation of temperature with heliocentric distance. The densities of Venus, Mars and the Earth are then due to different degrees of retention of S, O and H as FeS, FeO and hydrous silicates produced in chemical equilibrium

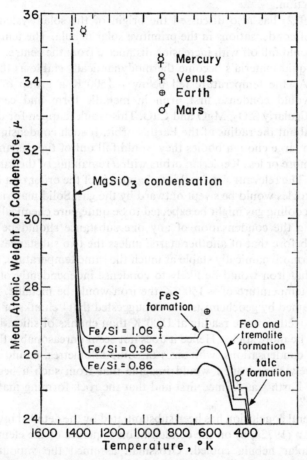

Fig. 2.4. Density of condensed material in equilibrium with a solar-composition gas, 400–1600°K at 10^{-3} bar. A simplified chemical system (the 20 most abundant elements) is employed for three different values of the Fe:Si ratio. The densities of some of the planets are in excellent agreement with an Fe:Si ratio of 1·08 but the omission of rare elements and uncertainties in the abundance of major elements could displace the entire manifold of curves slightly. (After Lewis, 1972b.)

between condensates and solar-composition gases. Only the Earth is likely to have differentiated so as to extract the heavy alkali metals into the core—thus the very large heat source ($\simeq 10^{20}$ erg/sec.), possible within the outer core of the Earth, is impossible on Mercury or Venus and unlikely on Mars. Lewis envisions the planets and smaller bodies to be formed by slow, low temperature accretion of condensate particles, initially by grain interaction forces ("stickiness", magnetic moments, electrostatic forces, etc.) but eventually, after growth of some protoplanets to multikilometer dimensions, by gravitational attraction.

Hoyle (1972) has also discussed the origin of the solar system from the viewpoint of condensations in the primitive solar nebula. The temperature of the gases would fall off with increasing distance d from the centre, and certain solid and liquid materials become thermodynamically stable as the temperature falls. As the temperature fell below $\sim 1500°K$ a group of refractory materials would condense first, iron in metallic form and certain metal oxides, particularly SiO_2, MgO and CaO. This would happen for $d \sim 2 \times 10^{13}$ cm, i.e. at about the radius of the Earth's orbit. If such condensing materials formed into large enough bodies they would fall out of the gas and continue moving in more or less Keplerian orbits with d remaining of the same order of magnitude. The relevant size for this to happen is of the order of a few metres. Smaller particles would be swept outward by the gas. Solid materials forming in a slowly cooling gas might be expected to be quite pure chemically. Also for slow cooling the condensation of any one substance should be essentially completed before that of another starts, unless the two substances happen to become thermodynamically stable at much the same temperature. Hoyle thus predicted that iron would be likely to condense independently of MgO and SiO_2. At a temperature of $\sim 1500°K$ the iron would be metallic, not FeS as usually assumed by geochemists. Hoyle suggested that "chunks of iron would aggregate together more readily at 1500°K than chunks of silica, because the iron would be more sticky. Hence it does not seem unreasonable that the first substantial condensations, with sizes of some kilometres, would be balls of comparatively pure iron. I would imagine that from such a beginning the core of the Earth was formed first and that the rock-forming materials were added later".

This general conclusion has been substantiated by the detailed investigations of Grossman (1972a, b) on the condensation sequence of the elements as the primitive solar nebula cooled. Grossman assumed the vapour to be in chemical equilibrium with each condensed phase over the entire temperature range between its condensation point and the temperature at which it was consumed by reaction to form new phases. Thus the derived sequence of condensation is an accurate description of the changing distribution of the elements between vapour and solid phase and between solid phases themselves since the effects of high temperature condensates on the composition of the gas were considered in determining the condensation temperatures of lower temperature species.

The temperature of condensation and disappearance of all phases (at 10^{-3} atm. total pressure) are given in Table 2.2. Metallic iron first appears at 1473°K and contains 12·1 mole per cent Ni. As the gas cools and more alloy

Table 2.2. Stability fields of equilibrium condensates at 10^{-3} atmospheres total pressure. (After Grossman 1972a.)

Phase		Condensation temperature (°K)	Temperature of disappearance (°K)
Corundum	Al_2O_3	1758	1513
Perovskite	$CaTiO_3$	1647	1393
Melilite	$Ca_2Al_2SiO_7$-$Ca_2MgSi_2O_7$	1625	1450
Spinel	$MgAl_2O_4$	1513	1362
Metallic Iron	(Fe, Ni)	1473	
Diopside	$CaMgSi_2O_6$	1450	
Forsterite	Mg_2SiO_4	1444	
	Ti_3O_5	1393	1125
Anorthite	$CaAl_2Si_2O_8$	1362	
Enstatite	$MgSiO_3$	1349	
Eskolaite	Cr_2O_3	1294	
Metallic Cobalt	Co	1274	
Alabandite	MnS	1139	
Rutile	TiO_2	1125*	
Alkali Feldspar	(Na, K)$AlSi_3O_8$	~1000	
Troilite	FeS	700	
Magnetite	Fe_3O_4	405	
Ice	H_2O	⩽200	

* Below this temperature, calculations were performed manually using extrapolated high temperature vapour composition data. In some cases, gaseous species which had been very rare assumed major importance at low temperature (CH_4).

condenses, its equilibrium Ni content decreases, reaching 4·9 mole per cent, corresponding to the Ni/Fe ratio of the solar system (Cameron, 1968), by 1350°K. Forsterite first appears at 1444°K, at which temperature 46 per cent of the total iron has already condensed.

The pressure dependence of the condensation temperatures of iron, forsterite and enstatite are shown in Fig. 2.5 from which it can be seen that iron has a higher condensation temperature than forsterite and enstatite at pressures above about $7·1 \times 10^{-5}$ and $2·5 \times 10^{-5}$ atm., respectively. The

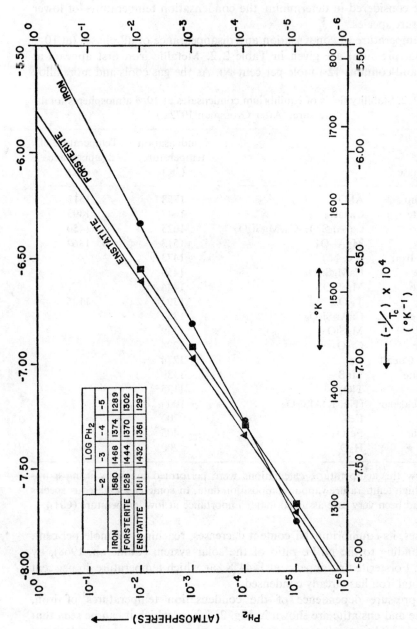

Fig. 2.5. The variation with pressure of the condensation temperatures of Fe, forsterite and enstatite. The depression of the condensation point of enstatite by the crystallization of forsterite has not been considered in this calculation. At any given pressure, the alloying of Ni in the metal will widen slightly the temperature difference between the appearance of Fe and forsterite. Although this temperature gap is small at 10⁻⁸ atm., 46 per cent of the Fe will have condensed before forsterite appears. (After Grossman, 1972b.)

difference between the temperatures of appearance of iron and forsterite gradually increases with increasing pressure, reaching approximately 80° at 10^{-1} atm. Forsterite condenses just before enstatite at all pressures above at least 10^{-6} atm., with the temperature gap between them also gradually increasing with increasing pressure. At 10^{-2} atm., forsterite condenses at 1528°K and enstatite at 1511°K. Equilibrium condensation of iron–nickel alloy proceeds in the same way over the entire pressure range investigated. The first alloy to condense is relatively nickel-rich but its nickel content rapidly decreases on cooling. Table 2.3 gives the compositions of the first condensing alloys and their condensation points at several different total pressures. The initial nickel content of the alloy appears to increase very

Table 2.3. Composition of the first condensing alloy, its condensation temperature and the fraction of the total iron condensed before the appearance of forsterite as a function of pressure. (After Grossman 1972a.)

Pressure (atm)	Condensation temperature of alloy (°K)	Initial nickel content of alloy (mole %)	Fraction of total iron condensed at condensation point of forsterite (%)
10^{-2}	1584	10·9	66
10^{-3}	1473	12·1	46
10^{-4}	1377	13·5	13
10^{-5}	1292	14·9	0
10^{-6}	1218	16·4	0

slowly with decreasing pressure, rising 1–1·5 mole per cent per ten-fold decrease in the total pressure.

In Grossman's inhomogeneous model, in a cooling gas of solar composition at 10^{-3} atm., total pressure, 46 per cent of the iron condenses before forsterite appears. As accretion continues, the release of gravitational potential energy causes melting of the core and the less dense components of the early condensate such as Al_2O_3, perovskite and melilite float to the surface. Below 1444°K, iron and magnesium silicates condense together and accrete upon the Ni–Fe planetary nucleus. Melting of silicates and iron takes place as a result of the impacting of infalling material and 5–10 per cent Si may enter the sinking metallic liquid by reactions such as those proposed by Brett (1971).

In Grossman's model, a large fraction of the total iron is buried inside the Earth at temperatures far above 700°K, below which it would have reacted with solar gases to form FeS. Although high temperature olivine and pyroxene are nearly Fe-free, severe loss of H_2 relative to H_2O by Jeans escape from the atmosphere of the planet allows 15–20 mole per cent of the iron end-members to enter these phases by the time the temperature has fallen to $1000°–900°K$. In the later stages of accretion, the accumulation rate decreases and the surface temperature falls, allowing more volatile and oxidized condensates to be retained by the Earth. Lower surface temperatures lead to a convective stage during which both the core and that part of the mantle already accreted are each internally homogenized.

A difficulty with the Turekian-Clark and Grossman models is that there does not appear to be enough energy available to melt the outer core either during or after accretion—it is even more difficult to do so for a cold accretion model. It was for this reason that Hanks and Anderson (1969) postulated a very short accretion time—less than 0·5 m yr. This difficulty can be alleviated to some extent if the lighter component of the core is sulphur. In an Earth of meteoritic composition, a sulphur-rich iron liquid would be the first melt to form (see Section 5.5). Core formation could proceed under these conditions at a temperature some 600°C lower than would be required to initiate melting in pure iron. In the vicinity of the core, the eutectic temperature is probably some 1600°C lower than that for pure iron (Hall and Murthy, 1972).

Anderson and Hanks (1972) have reconsidered the inhomogeneous accretion model and concluded that it could account for the early melting of the core. In their model the proto-Earth consists of a uranium/thorium rich central nucleus, composed chiefly of Ca, Al and Ti rich silicates; a shell of Fe–Ni containing some of the earlier condensates which had not fully con-densed or accreted when the Fe and Ni condensed; a shell of less refractory silicates, mainly pyroxene and olivine; a shell of potassium and sodium rich silicates; and finally a shell of hydrated minerals and volatile rich condensates. The proto-core is composed of the refractory nucleus and the Fe–Ni shell. It is the presence of a radioactive nucleus that provides the mechanism for melting the metal shell. In the model of Anderson and Hanks, melting of Fe commences at about $0·4 \times 10^9$ yr. The nucleus and the proto-core are not in gravitational equilibrium because Fe is denser than the nucleus. As melting of Fe commences, the nucleus will attempt to rise—the rise of the nucleus to the base of the mantle will not be symmetric because of the convective pattern in the core which is controlled by the rotation of the Earth. When the nucleus leaves the centre of the Earth it will be replaced by nickel–iron which, before its descent, is close to its melting point. Because of the steep slope of the melting curve relative to the adiabatic curve, it will refreeze, explaining the seismic evidence of the solidity and composition of the inner core. Their model also suggests that part of the extra light alloying material in the molten outer core may be residue from the nucleus, i.e. Ca–Al rich oxides and silicates either in solution or in suspension. In this case, the high radioactivity of the material could provide part of the energy for driving the geomagnetic dynamo (see Section 4.7).

Many of the questions raised in this chapter will be discussed again in Chapter 5 on the constitution of the Earth's core. Grossman and Larimer (1974) have recently reviewed the literature on chemical fractionations during the condensation of the solar system and their consequences on the establishment of chemical differences between the different classes of chondrites and between the planets.

2.6. Variation of the Gravitational Constant *G* with Time

Another physical process that could affect the evolution of the Earth is the possibility that the gravitational constant G varies with time. A number of geophysicists e.g. Egyed (1956), Carey (1958), Heezen (1959), Wilson (1960) have suggested that the Earth has been expanding with time. In Carey's hypothesis the expansion took place mainly during the past 500 m yr leading to an average rate of increase in the radius of the Earth during this period of about 5 mm/yr; Egyed inferred a rate of increase of the Earth's radius of 0·4–0·8 mm/yr. One of the mechanisms which has been suggested to account for such an expansion is Dirac's (1938) speculation that the gravitational constant G varies inversely with time. With a gradual decrease of G, the pressure would decrease inside the Earth and the volume increase. Dicke (1957) showed that to account for the expansion demanded by Carey by this mechanism it would be necessary to assume that G has been decreasing at a rate of roughly 1 part in 10^8 per yr: to meet the expansion of Egyed, the rate of decrease of G would have to be 1 part in 10^9 per yr.

Egyed (1960) first suggested that palaeomagnetic data might be used to calculate the radius of the Earth at various times in the past. On the expansion hypothesis of Carey, Heezen, and Egyed, the continents do not increase in area, so that the distance between any two points on a stable part of one continent remains the same. Thus if the Earth's radius increases, the geocentric angle between the two points decreases. Assuming the Earth's ancient magnetic field to be dipolar, the Earth's ancient radius may then be found from the measured inclinations of contemporaneous rocks from two localities on the same stable continental block. Cox and Doell (1961) used this method on Permian data from Europe and Siberia to obtain an estimated Permian radius of 0·99 times the present radius. Ward (1963) generalized Egyed's method of calculation and applied it to Devonian and Triassic data as well as Permian from Europe and Siberia. He obtained estimated Earth radii for these periods of 1·12, 0·94 and 0·99 times the present radius respectively and considered none of these estimates to be significantly different from the present radius.

Using high pressure shock wave data, Birch (1968) showed that the increase in radius of an Earth having a chemically distinct mantle and core would only be about 370 km for a decrease in the gravitational constant from $2G$ to its present value of G. A larger increase in radius would be possible if Ramsey's hypothesis (see Section 5.2) were true—Birch showed, however, that Ramsey's hypothesis is extremely unlikely to hold. He concluded that if the mass of the Earth has remained constant, changes in the Earth's radius are unlikely to exceed 100 km. Thus, the sum total of evidence indicates that any large expansion of the Earth has not taken place and that the upper limit to any rate of change of G is about 1 part in 10^{10} per yr.

If G has been decreasing with time, the rate of radiation of the sun would have been higher in the past and hence asteroids and meteorite bodies would have been warmer, possibly leading to loss of argon from the material of the meteorites. From the observed K–Ar ages of meteorites, Peebles and Dicke (1962) have shown that G cannot have been decreasing by more than about 1 part in 10^{10} per yr. This rules out the possibility that a decrease in G could lead to the large expansions required by Carey and Egyed: the limit of 1 part in 10^{10} per yr in the variation in G leads to an upper limit in the rate of increase in the Earth's radius of about 0·05 mm/yr. Further deductions by Dicke (1966) on various geophysical effects indicate that only a very small decrease in G is possible. On the other hand, Hoyle (1972) is in favour of a decrease in G with time. One result of this would be higher temperatures in the past—the mean sea level temperature of the Earth being about 70°C 2000 m yr ago. The most serious objection to a variable G is, in Hoyle's opinion, the Precambrian glaciation of the Canadian shield which has been estimated to have occurred 2500 m yr ago.

Recently there has been a fairly rapid increase in the number of observations of "discrepant redshifts". Hoyle and Narlikar (1971) have given a possible explanation in a theory which incorporates a gravitational constant G that is decreasing with time. In a later paper (1972) they showed that their theory implies that the radius of the Earth has increased at a rate of about 0·1 mm/yr. Shapiro et al. (1971) have placed an observational upper limit of 4×10^{-10}/yr on $\dot{G}/G$. Taking Hubble's constant as 5×10^{-11} yr, the variation expected would be 10^{-10}/yr which is in close agreement with Dicke's (1962) estimate.

If G decreases by a sufficient amount a cluster of particles will expand. In some cases the particles may have sufficient velocity to escape from the cluster as the gravitational binding decreases. Since it is known that clusters of galaxies and globular clusters currently exist with finite dimensions, Dearborn and Schramm (1974) were able to set limits on the magnitude of the variation of G, by investigating numerically the dynamical effects on an isolated cluster of galaxies caused by non zero values of $\dot{G}/G$. For a range of initial conditions, the maximum rate of change in G was determined which would still allow the existence of clusters of galaxies at the current epoch. A similar study was made for globular clusters and the results were found to be comparable. They found an upper limit for $\dot{G}/G$ of 4–10×10^{-11}/yr. This limit is much stronger than that of Shapiro et al. (1971). However the Dirac cosmology is not consistent with this limit. Also the Hoyle–Narlikar (1972) cosmology with $\dot{G}/G \propto 1/t$ would seem to be inconsistent with the observations. On the other hand Morrison (1973) obtained a limit of similar magnitude to that given by Dearborn and Schramm. He used occultations timed on atomic scales to determine the slowing of the Earth's rotation due to tidal and nontidal (i.e. depending on $\dot{G}$) forces. Previous determinations were based on data timed on a gravitational basis on which the assumed effects of $\dot{G}/G$ would not be apparent, leaving only tidal slowing. The close agreement of the deceleration rates then implies $\dot{G}/G=0$, yielding an upper limit of 2×10^{-11}/yr on $\dot{G}/G$ from the uncertainty in the data. This result is approximately the same as that of Dearborn and Schramm.

There have been a number of suggestions for obtaining information about the strength of the Earth's gravitational field in the past. To obtain mean rates of change over millions of years a gravity sensitive geological system is required. There is no shortage of gravity controlled phenomena but gravity seldom leaves a perment record in the rocks and the effects of gravity are usually small compared with those of other often unpredictable variables.

Stewart (1970) has suggested a number of phenomena, both geological and biological, which could possibly be used for this purpose. These include the gravitational compaction of clays, the compaction of clays beneath glacial ice, palaeobarometry, the size of flying animals, the depth of animals' footprints, the dimensions of the frames of land animals, and the growth of diapirs. Later (Stewart, 1972) he developed a method which, while not determining accurate values of palaeogravity, has been able to define a limit to the decrease of g with time. The limit, though broad, is not inconsistent with theoretical predictions. Stewart argued that if gravity in the past had been higher than today it is conceivable that some fine grained sedimentary deposits would be over consolidated, i.e. compacted more by the smaller sedimentary column above them in the past than by the larger one existing today. Since compaction is relatively rapid and largely irreversible the requisite evidence could be preserved. Stewart measured the degree of over-consolidation in sediments in the London basin. He found that the London clay now exposed has been consolidated by higher pressures than would have been produced by what is now the greatest thickness of sedimentary overburden to be found anywhere in the London basin. It is of course possible that this extra pressure was derived from sediments younger than any now observed, additional overburden which has since been eroded away. However if the additional pressure was caused solely by a higher value of g in the past (about 26 m yr ago), Stewart's results indicate that gravity at that time could not have been more than twice its present value. This implies that the maximum possible decrease in g over the past 26 m yr is 4 parts in 10^8 per yr.

References

Alfvén, H. and Arrhenius, G. (1970a). Structure and evolutionary history of the solar system. I. *Astrophys. Space Sci.* **8**, 338.

Alfvén, H. and Arrhenius, G. (1970b). Origin and evolution of the solar system, II. *Astrophys. Space Sci.* **9**, 3.

Alfvén, H. and Arrhenius, G. (1973). Structure and evolutionary history of the solar system, III. *Astrophys. Space Sci.* **21**, 117.

Anders, E. (1968). Chemical processes in the early solar system, as inferred from meteorites. *Acc. Chem. Res.* **1**, 289.

Anderson, D. L. and Hanks, T. C. (1972). Formation of the Earth's core. *Nature* **237**, 387.

Arrhenius, G., Asunmaa, S. K. and Fitzgerald, R. W. (1972). *In* Lunar Sci. Inst. Contribution No. 88 in Lunar Science III (Ed. C. Watkins).

Birch, F. (1965). Energetics of core formation. *J. geophys. Res.* **70**, 6217.

Birch, F. (1968). On the possibility of large changes in the Earth's volume. *Phys. Earth Planet. Int.* **1**, 141.

Blander, M. and Katz, J. L. (1967). Condensation of primordial dust. *Geochim. cosmochim. Acta* **31**, 1025.

Blander, M. and Abdel-Gawad, M. (1969). The origin of meteorites and the constrained equilibrium condensation theory. *Geochim. cosmochim. Acta* **33**, 701.

Brecher, A. (1971). On the primordial condensation and accretion environment and the remanent magnetization of meteorites. *In* "Proceedings of the IAU Symposium on the Evolutionary and Physical Properties of Meteoroids", Albany, June, 1971.

Brett, R. (1971). The Earth's core; speculations on its chemical equilibrium with the mantle. *Geochim. cosmochim. Acta* **35**, 203.

Cameron, A. G. W. (1968). A new table of abundances of the elements in the solar system. *In* "Origin and Distribution of the Elements", (Ed. L. H. Ahrens), Pergamon, Oxford.

Cameron, A. G. W. (1972). Models of the primitive solar nebula. *In* "Symp. Origin of the Solar System", Nice 1972 (Ed. H. Reeves), Cent. Nat. Rech. Scient., Paris.

Carey, S. W. (1958). A tectonic approach to continental drift. *In* "Continental Drift, a Symposium", Univ. Tasmania, Hobart.

Chandrasekhar, S. (1961). "Hydrodynamic and Hydromagnetic Stability", Oxford University Press, London.

Clark, S. P., Jr., Turekian, K. K. and Grossman, L. (1972). Model for the early history of the Earth. *In* "The Nature of the Solid Earth" (Ed. E. C. Robertson), McGraw-Hill, New York.

Cox, A. and Doell, R. R. (1961). Palaeomagnetic evidence relevant to a change in the Earth's radius. *Nature* **189**, 45.

Deerborn, D. S. and Schramm, D. N. (1974). Limits on variation of *G* from clusters of galaxies. *Nature* **247**, 441.

Dicke, R. H. (1957). Principle of equivalence and the weak interactions. *Rev. mod. Phys.* **29**, 355.

Dicke, R. H. (1962). Implication for cosmology of stellar and galactic evolution rates. *Rev. mod. Phys.* **34**, 110.

Dicke, R. H. (1966). The secular acceleration of the Earth's rotation and cosmology. *In* "The Earth–Moon System" (Ed. B. G. Marsden and A. G. W. Cameron), Plenum Press, New York.

Dirac, P. A. M. (1938). A new basis for cosmology. *Proc. R. Soc.* A **165**, 199.

Egyed, L. (1956). A new theory on the internal constitution of the Earth and its geological–geophysical consequences. *Acta. geol. hung.* **4**, 43.

Egyed, L. (1960). Some remarks on continental drift. *Geofis. pura appl.* **45**, 115.

Elsasser, W. M. (1963). Early history of the Earth. *In* "Earth Science and Meteoritics" (Ed. J. Geiss and E. D. Goldberg), North-Holland Publ. Co., Amsterdam.

Eucken, A. (1944). Physikalisch-chemische Betrachtungen über die früheste Entwicklungsgeschichte der Erde, Nachr Akad. Wiss. Göttingen, Math-Phys. Kl., Heft 1, 1.

Fireman, E. L., DeFelice J. and Norton, E. (1970). Ages of the Allende meteorite. *Geochim. cosmochim. Acta* **34**, 873.

Fish, R. A., Goles, G. G. and Anders, E. (1960). The record in the meteorites. III: On the development of meteorites in asteroidal bodies. *Astrophys. J.* **132**, 243.

Flasar, F. M. and Birch, F. (1973). Energetics of core formation: a correction. *J. geophys. Res.* **78**, 6101.

Fleischer, R. L. and Naeser, C. W. (1972). Search for plutonium-244 tracks in Mountain Pass bastnaesite. *Nature* **240**, 465.

Gastil, G. (1960). The distribution of mineral dates in time and space. *Am. J. Sci.* **258**, 1.

Gray, C. M. and Compston, W. (1974). Excess ^{26}Mg in the Allende meteorite. *Nature* **251**, 495.

Grossman, L. (1972a). Condensation in the primitive solar nebula. *Geochim. cosmochim Acta* **36**, 597.

Grossman, L. (1972b). Condensation, chondrites and planets. Ph.D. thesis, Yale University.

Grossman, L. and Larimer, J. W. (1974). Early chemical history of the solar system. *Rev. Geophys. Space Phys.* **12**, 71.

Hall, H. T. and Murthy, V. R. (1972). Comments on the chemical structure of an Fe–N–S core of the Earth, Int. Conf. Core–Mantle Interface, *Trans. Am geophys. Un.* **53**, 602.

Hanks, T. C. and Anderson, D. L. (1969). The early thermal history of the Earth. *Phys. Earth Planet. Int.* **2**, 19.

Harris, P. G. and Tozer, D. C. (1967). Fractionation of iron in the solar system. *Nature* **215**, 1449.

Heezen, B. C. (1959). Palaeomagnetism, continental displacements and the origin of submarine topology. *Inter. Ocean. Cong.*, 1959.

Hoffman, D. C., Lawrence, F. O., Mewherter, J. L. and Rourke, R. M. (1971). Detection of Plutonium-244 in Nature. *Nature* **234**, 132.

Hoyle, F. (1972). The history of the Earth. *Q. Jl. R. astr. Soc.* **13**, 328.

Hoyle, F. and Wickramasinghe, N. C. (1968). Condensation of the planets. *Nature* **217**, 415.

Hoyle, F. and Narlikar, J. V. (1971). On the nature of mass. *Nature* **233**, 41.

Hoyle, F. and Narlikar, J. V. (1972). Cosmological models in a conformally invariant gravitational theory. II: A new model. *Mon. Not. R. astr. Soc.* **155**, 323.

Kerridge, J. F. and Vedder, J. F. (1972). Accretionary processes in the early solar system: an experimental approach. *Science* **177**, 161.

Larimer, J. W. (1967). Chemical fractionation in meteorites. I: Condensation of the elements. *Geochim. cosmochim. Acta* **31**, 1215.

Larimer, J. W. and Anders, E. (1970). Chemical fractionation in meteorites. III: Major element fractionations in chondrites. *Geochim. cosmochim. Acta* **34**, 367.

Lee, T. and Papanastassiou, D. A. (1974). Mg isotopic anomalies in the Allende meteorite and correlation with O and Sr effects. *Geophys. Res. Letters* **1**, 225.

Lewis, J. S. (1972a). Low temperature condensation from the solar nebula. *Icarus* **16**, 241.

Lewis, J. S. (1972b). Metal/silicate fractionation in the solar system. *Earth Planet. Sci. Letters* **15**, 286.

Lewis, J. S. (1973). Chemistry of the planets. *A. Rev. phys. Chem.* **24**, 339.

MacDonald, G. J. F. (1959). Calculations on the thermal history of the Earth. *J. geophys. Res.* **64**, 1967.

Maurette, M. and Bibring, J. P. (1972). Stellar wind radiation damage in cosmic dust grains: implications for the history of early accretion in the solar nebula. *In* "Symp. Origin of the Solar System", Nice 1972 (Ed. H. Reeves), Cent. Nat. Rech. Scient., Paris.

Morrison, L. V. (1973). Rotation of the Earth from A.D. 1663–1972 and the constancy of *G*. *Nature* **241**, 519.

Neukum, G. (1968). Ph.D. Thesis, University of Heidelberg.

Orowan, E. (1969). Density of the moon and nucleation of planets. *Nature* **222**, 867.

Oversby, V. M. and Ringwood, A. E. (1971). Time of formation of the Earth's core. *Nature* **234**, 463.

Patterson, C. and Tatsumoto, M. (1964). The significance of lead isotopes in detrital feldspar with respect to chemical differentiation within the Earth's mantle. *Geochim. cosmochim. Acta* **28**, 1.

Peebles, J. and Dicke, R. H. (1962). The temperature of meteorites, Dirac's cosmology and Mach's Principle. *J. geophys. Res.* **67**, 4063.

Podosek, F. A. (1970). Dating of meteorites by the high-temperature release of iodine-correlated Xe^{129}. *Geochim. cosmochim. Acta* **34**, 341.

Podosek, F. A. and Lewis, R. S. (1972). ^{129}I and ^{244}Pu abundances in white inclusions of the Allende meteorite. *Earth Planet. Sci. Letters* **15**, 101.

Purcell, E. M. and Spitzer, L. (1971). Orientation of rotating grains. *Astrophys. J.* **167**, 31.

Rao, M. N. and Gopalan, K. (1973). Curium-248 in the early solar system. *Nature* **245**, 304.

Reeves, H. (Ed.) (1972). "Symposium on the Origin of the Solar System", Nice 1972, Cent. Nat. Rech. Scient., Paris.

Ringwood, A. E. (1959). On the chemical evolution and densities of the planets. *Geochim. cosmochim. Acta* **15**, 257.

Ringwood, A. E. (1960). Some aspects of the thermal evolution of the Earth. *Geochim. cosmochim. Acta* **20**, 241.

Ringwood, A. E. (1966). Chemical evolution of the terrestrial planets. *Geochim. cosmochim. Acta* **30**, 41.

Rowe, M. W. and Kuroda, P. K. (1965). Fissiogenic xenon from the Passamonte meteorite. *J. geophys. Res.* **70**, 709.

Runcorn, S. K. (1962a). Towards a theory of continental drift. *Nature* **193**, 311.

Runcorn, S. K. (1962b). Palaeomagnetic evidence for continental drift and its geophysical cause. *In* "Continental Drift" (Ed. S. K. Runcorn), Academic Press, New York.

Safronov, V. S. (1969). Evolution of the protoplanetary cloud and formation of the Earth and planets. *Nauka, Mowcow*, (1969); translated *NASA TTF*-677, (1972).

Safronov, V. S. (1972). Accumulation of the planets. *In* "Symp. Origin of the Solar System", Nice 1972 (Ed. H. Reeves), Cent. Nat. Rech. Scient., Paris.

Sakamoto, K. (1974). Possible cosmic dust origin of terrestrial plutonium-244. *Nature* **248**, 130.

Schramm, D. M., Tera, F. and Wasserburg, G. J. (1970). The isotope abundance of ^{26}Mg and limits on ^{26}Al in the early solar system. *Earth Planet. Sci. Letters* **10**, 44.

Shapiro, I. I., Smith, W. B., Ash, M. B., Ingalls, R. P. and Pettengill, G. H. (1971). Gravitational constant: experimental bound on its time variation. *Phys. Rev. Letters* **26**, 27.

Stewart, A. D. (1970). Palaeogravity. *In* "Palaeogeophysics" (Ed. S. K. Runcorn), Academic Press, London.

Stewart, A. D. (1972). Palaeogravity from the compaction of fine-grained sediments. *Nature* **235**, 322.

Tolland, H. G. (1973). Formation of the Earth's core. *Nature Phys. Sci.* **243**, 141.

Tozer, D. C. (1965). Thermal history of the Earth. 1: The formation of the core. *Geophys. J.* **9**, 95.

Tozer, D. C. (1972). The present thermal state of the terrestrial planets. *Phys. Earth Planet. Int.* **6**, 182.

Turekian, K. K. and Clark, S. P., Jr. (1969). Inhomogeneous accumulation of the Earth from the primitive solar nebula. *Earth Planet. Sci. Letters* **6**, 346.

Urey, H. C. (1952). "The Planets", Yale University Press.

Urey, H. C. (1962). The origin of the moon and its relationship to the origin of the solar system. *In* "The Moon" (Proc. I.A.U. 14th Symp., Ed. Z. Kopal and Z. K. Mikhailov), Academic Press, New York.

Vening Meinesz, F. A. (1952). Convection currents in the Earth and the origin of the continents I. *Kon. Ned. Akad. Weten.* **55**, 527.

Ward, M. A. (1963). On detecting changes in the Earth's radius. *Geophys. J.* **8**, 217

Wetherill, G. W. (1972). The beginning of continental evolution. *Tectonophysics* **13**, 31.

Wilson, J. T. (1960). Some consequences of expansion of the Earth. *Nature* **185**, 880.

3. The Thermal Regime of the Earth's Core

3.1. Introduction

Of all the properties of the Earth's interior, its temperature is the least well known. It depends upon the initial temperature (and hence on the origin of the Earth), the distribution of radioactive elements, and the mode of heat transfer in the Earth. In this last respect convection and the radiative transfer of heat must be considered in addition to lattice conduction. A fundamental problem in the thermal history of the Earth is an explanation of the present regime—a solid inner core and mantle and a liquid outer core. This will be considered in the next section.

3.2. The Earth's Inner Core

The seismological evidence for a sharp inner core boundary and possibly one or more first-order discontinuities in velocity at greater radii raises the important question of possible liquid–solid or solid–solid phase transitions in either a homogeneous material or in a multi-component chemical system. If the transition from the inner to the outer core is a transition from the solid to the liquid form of a single material, then the boundary must be at the melting point and a constraint is put on the thermal regime of the Earth's interior. Jacobs (1953a) used this fact to explain how the mantle and inner core could be solid, while at the same time the outer core is liquid.

Consider the simplest type of model first—a core of pure iron and a mantle of silicates. At the boundary between the mantle and core there will then be a discontinuity in the melting-point–depth curve, although the actual temperature must be continuous across the boundary. The form of this discontinuity

89

D

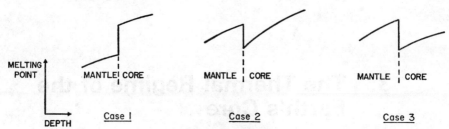

Fig. 3.1. Possible forms of the melting-point–depth curve in the neighbourhood of the mantle–core boundary.

could, mathematically, take any of the three cases shown in Fig. 3.1. Case 1, in which the melting-point curve in the core is always above that in the mantle, is impossible; for the actual temperature curve must lie below the melting-point curve in the mantle, above it in the core, and yet be continuous across the boundary. Cases 2 and 3 are both possible. In case 3, the melting-point curve in the core never rises above the value of the melting point in the mantle at the MCB, while in case 2 it exceeds this value for part of the core. Considering first case 2, the melting-point curve will be of the general shape shown in Fig. 3.2.

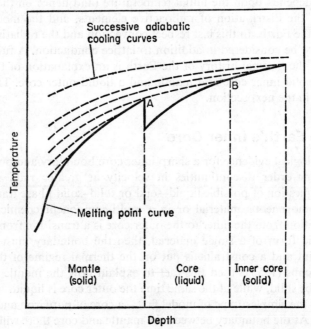

Fig. 3.2. Melting-point curve and successive adiabats in the Earth's interior. (After Jacobs, 1953a.)

If the Earth cooled from a molten state, the temperature gradient would be essentially adiabatic, there being strong convection currents and rapid cooling at the surface. Solidification would commence at that depth at which the curve representing the adiabatic temperature first intersected the curve representing the melting-point temperature. Solidification would thus begin at the centre of the Earth. A solid inner core would then continue to grow until a curve representing the adiabatic temperature intersected the melting-point curve twice, one at A, the boundary between the core and mantle, and again at B, as shown in Fig. 3.2. As the Earth cooled still further, the mantle would begin to solidify from the bottom upward. The liquid layer between A and B would thus be trapped. The mantle would cool at a relatively rapid rate, leaving this liquid essentially at its original temperature, insulated above by a rapidly thickening shell of silicates and below by the already solid (iron) inner core.

In the above discussion no specific values of the temperatures are postulated, and the behaviour of the adiabatic and melting-point curves need not be known exactly. If they vary qualitatively as shown, the above argument gives a physical explanation for the existence of a solid inner core. It follows by similar reasoning that if the melting-point–depth curve in the neighbourhood of the MCB is as shown in case 3 of Fig. 3.1, then as the Earth cooled from a molten state, the entire core would be left liquid. The physical state of the inner core thus depends on the magnitude of the discontinuity in the melting-point curve at the MCB. Finally, if the Earth had a cold origin and never became completely molten, then as the temperature increased with time either case 2 or case 3 could lead eventually to a liquid outer core with a solid inner core. However if the Earth were never completely molten, it is difficult to explain the differentiation into core and mantle and hence a melting-point–depth curve of the form shown in Fig. 3.2. The Earth would have had to have accreted inhomogeneously (see Section 2.5).

The above arguments break down if the melting point and adiabatic temperature curves of Higgins and Kennedy (1971) are correct—for in their curves the adiabatic gradient is steeper than the melting point gradient in the outer core. The shape of these two curves is discussed in detail in the next two sections.

3.3. Melting-Point–Depth Curves

Simon (1937) formulated a semi-empirical equation for the dependence of the melting-point on pressure which has often been applied to geophysical problems. Simon's equation is

$$p = A\left[\left(\frac{T_m}{T_{m_0}}\right)^c - 1\right] \qquad (3.1)$$

where A is a constant, related to the internal pressure, T_m the melting-point temperature at pressure p, T_{m_0} the melting point at atmospheric pressure, and c a numerical constant. A partial theoretical basis for Simon's equation has been given by Salter (1954) and Gilvarry (1956a, b, c). Gilvarry showed that the constant c is related to Grüneisen's parameter γ through the equation

$$c = \frac{6\gamma + 1}{6\gamma - 2}. \tag{3.2}$$

The Simon equation has been applied to the Earth's core on the assumption that it is pure iron (Simon, 1953; Jacobs, 1953a; Gilvarry, 1957); Bullard (1954) has also used it to estimate melting temperatures in both the mantle and the core.

Another empirical expression for the melting-point at high pressures has been proposed by Kraut and Kennedy (1966). They showed that the Simon equation for metals almost invariably predicts higher melting temperatures at higher pressures than are measured in the laboratory, and that for most metals there is a linear relationship between T_m and the fractional change in volume $(\Delta V/V_0)_s$ of the solid resulting from the applied pressure. Their equation is

$$T_m = T_{m_0}\left\{1 + c\left(\frac{\Delta V}{V_0}\right)_s\right\} \tag{3.3}$$

where c is a constant depending on the particular substance. Substances bonded by van der Waals' forces yield curves concave to the temperature axis; a small number of ionically bonded substances to which the relationship

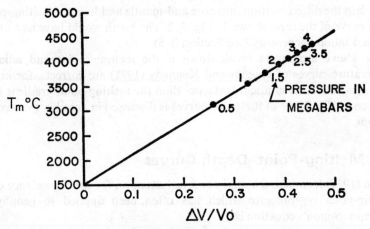

Fig. 3.3. Melting temperature T_m of iron versus fractional change in volume $\Delta V/V_0$. (After Higgins and Kennedy, 1971.)

has been applied give curves convex to the temperature axis. Kennedy and Vaidya (1970) showed however that materials with metallic bonding result in general in a straight line relationship. Gilvarry (1966) has demonstrated that Eqn. (3.3) may be derived from the Lindemann law as reformulated by himself (1956a, 1957). He also showed that it is equivalent, in the special case of a restricted range of melting temperatures, to a relationship in which the constant c appears in terms of the Grüneisen parameter of the solid at the normal fusion point. Gilvarry maintains that Eqn. (3.3) is thus no more than an interpolation (or extrapolation) in the sense established for the Simon equation (Gilvarry, 1956c), and is not a fundamental criterion of melting.

If the linear relationship (Eqn. 3.3) between T_m and $(\Delta V/V_0)_s$ holds for iron, a melting-point–depth curve may be obtained for the core of the Earth provided the initial slope of the melting curve of iron is known and also the relationship between pressure and the specific volume of iron along its melting curve. Such data have been provided by the experimental work on the melting of iron by Sterrett *et al.* (1965) and by the results of shockwave experiments on the density of iron at high pressures (Van Thiel, 1966). Higgins and Kennedy (1971) have thus re-estimated the melting-point gradient in the core of the Earth. Figure 3.3 shows their linear relationship between T_m and $(\Delta V/V_0)_s$ with the various pressure intervals marked on the curve, and Fig. 3.4 gives their plot of the melting curve of iron versus depth in the Earth. Their melting-point gradient in the Earth's core is much less steep than earlier estimates—there being an increase of only 500°C in the melting-point across the outer core. The effect of pressure on the melting temperature of metals

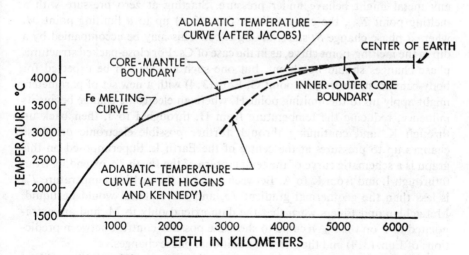

Fig. 3.4. Melting-point curve of iron and adiabatic temperature curves of Jacobs (1971b) and Higgins and Kennedy (1971) versus depth in the Earth's interior.

has been further investigated by McLachlan and Ehlers (1971), based on the theory that an entrapped gaseous fraction is formed when a metal changes from the solid to the liquid state. Their equation is

$$T_m = T_{m_0} + \frac{V_0}{\beta' R} \left(\frac{\Delta V}{V_0} \right) \{1 - \exp(-\beta' p)\} \tag{3.4}$$

where R is the gas constant and β' an "effective" compressibility constant. The exponential term allows for the (observed) decrease in slope dT_m/dp with increase in pressure. Equation (3.4) is based primarily on kinetic theory and is in good agreement with experimental curves of (p, T_m) obtained for Li, Na, K, Rb, Fe and Ni. McLachlan and Ehlers warn however against extrapolating the melting curve for Fe to pressures at the MCB because of our lack of knowledge of possible electronic and phase changes that may occur at such high pressures. One metal, Cs, has been subjected to pressures high enough to cause both types of changes to occur (Jayaraman et al., 1967). Cs shows a known structural phase change between 20 and 30 kbar and is accompanied by a mild, but distinct, dip in the melting-point curve. At higher pressures a phase change occurs whose cause was attributed by Fermi to electronic collapse of the atoms. The electronic phase change occurs first in the liquid over a broad pressure range. This causes a large drop in the melting-point curve to 42·5 kbar where the solid undergoes an abrupt volume change of about 9 per cent, followed by a large change in slope of the melting-point curve.

Figure 3.5, after McLachlan and Ehlers (1971), shows schematically how any metal might behave under pressure. Starting at zero pressure with a melting point T_{m_0}, the curve follows Eqn. (3.4) up to a limiting point M, where a phase change St starts taking effect. This may be accompanied by a dip in the melting-point curve, as in the case of Cs. For close-packed structures phase changes should be absent, but one such change can be expected for body-centred structures. At point G, Eqn. (3.4) with a new set of parameters might apply up to the limiting point H, where an electronic change begins its influence, reducing the temperature from H, through I to J, then back up through K, and continuing through further possible electronic or phase changes up to pressures at the centre of the Earth L. Superimposed on this graph is a schematic curve of the temperature of the Earth Te, extending from 0 through I, and from K to A. Between I and K the melting temperature T_m is less than the geothermal gradient Te, and the material would be liquid. McLachlan and Ehlers warn that the data extend only to M, and are extrapolated to L on the T_m–p curve to show the possible contrast between predictions of Eqn. (3.4) and the effect of additional phase changes.

Birch (1972) has pointed out that most discussions of the melting of iron neglect the fact that iron exists in several cyrstalline forms and that the melting curve defines a condition of equilibrium between the liquid phase and one of the solid phases. Four crystalline phases of iron are known (α, γ, δ and ϵ). Only at low pressures is the α phase in equilibrium with the melt. Our knowledge of the thermodynamic properties of the γ and ϵ-phases (of most geophysical interest) is very rudimentary. Birch estimated that the γ melting temperatures are about 700° higher than those of Higgins and Kennedy, with ϵ melting temperatures still higher; he does not believe that present evidence

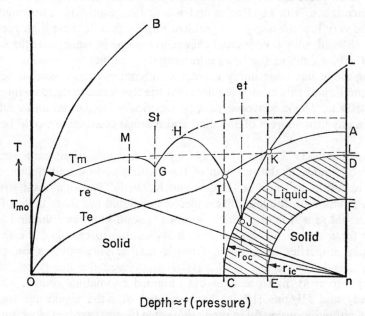

Fig. 3.5. Hypothetical extension of melting-point data to the Earth's core. The coordinates represent change of p and T with depth. Curve Tm shows schematic changes of the melting-point curve, and Te is a possible geothermal gradient. The point K coincides with the inner–outer core boundary, whereas I may either be at the outer core–mantle boundary or within the lower mantle. (After McLachlan and Ehlers, 1971.)

is sufficient to predict melting temperatures of iron at core pressures to within 500°.

Verhoogen (1973) has also discussed the melting of iron pointing out that the recent estimate of Higgins and Kennedy is based on a hundred-fold linear extrapolation of the 30 kbar experimental results of Sterrett *et al.* (1965). These measurements are themselves somewhat uncertain as they also require an extrapolation, over a range of several hundred degrees, of the effect of pressure on the thermocouples. Higgins and Kennedy extrapolated results from 30 kbar to 3·3 Mbar, by assuming that the melting-point is a linear function of the fractional change in volume of the solid. This law, (Kraut and Kennedy, 1966) which is known not to be valid for non metals, is assumed to hold for all metals including iron, even though there are metals to which it obviously does not apply such as those which have a negative initial slope of the melting-point curve (e.g. Bi), or those (e.g. Cs) which have one or several maxima in the melting-point curve. Also, as already mentioned, Higgins and Kennedy ignore the phase changes ($\delta \rightarrow \gamma$ and $\gamma \rightarrow \epsilon$) which occur in iron and which must affect the slope of the melting-point curve. Verhoogen thus does not place much faith in Higgins and Kennedy's results. He also points out that the very low melting-point gradient implies that at these high pressures iron melts with only a very small change in volume in which case the density jump (~ 0.5 g/cm^3) at the inner core boundary cannot be accounted for by melting alone but must imply a notable difference in composition between solid and liquid: this in turn requires that the temperature there be much less than 4500°K. The core would then be practically isothermal and could not contain any heat sources of any kind and thermal convection would be ruled out.

Leppaluoto (1972) has recently applied Eyring's significant structure theory of liquids to the problem of the melting of iron and obtained very different results to those of Higgins and Kennedy. The significant structure theory is based on the idea that molecules of liquid are more or less free to move about in a structure which is basically solid-like (see John and Eyring (1971) for a review on the subject). It is possible to calculate the thermodynamic properties of a liquid from a mathematical expression of this concept, and Tuerpe and Keeler (1967) have used the theory to estimate melting curves at high pressures but obtained anomalous results. Thus, as Kennedy and Higgins (1973a) have pointed out, the significant structure theory, although successful in predicting some thermal properties of liquids at moderate temperatures and pressures, may not be applicable to describe the phenomenon of melting. In Leppaluoto's study the melting-point at high pressures is determined by equating the free energies of liquid and solid: shock wave data are used to determine the volume and other properties of the solid. The method has the advantage of removing some of the empiricism attached to other current melting laws which focus exclusively on the properties of the solid and ignore the essence of the melting phenomenon as a two-phase equilibrium. Leppaluoto's calculations however involve a parameter (the activation volume) which is not well known so that at very high pressures the melting-point can only be estimated to lie within a certain range. The melting-point predicted by Higgins and Kennedy corresponds to assuming the activation volume to be zero: this assumption, above 100 kbar, leads to thermodynamically inconsistent results. Leppaluoto estimated the melting-point for iron to be $\gtrsim 5000$°K at the MCB (≈ 1.4 Mbar) and $\gtrsim 7000$°K at the inner core–outer core boundary ($\approx 3\cdot3$ Mbar). Alder (1966), using the Lindemann law, where melting is assumed to occur when the root-mean-square amplitude of atomic vibrations becomes some critical fraction of the nearest-neighbour distance, estimated the melting temperature of iron under MCB conditions to be about 4400°C, and under inner core–outer core boundary conditions to be about 7450°C.

Boschi (1974a) has investigated the asymptotic behaviour of the melting curves for substances with close packed structures by means of Monte Carlo calculations on model systems of hard spheres. A review of this method of estimating equilibrium averages of variables of general interest in statistical mechanics and thermodynamics has been given by Ree (1971). Boschi's results, although consistent with Simon's empirical equation (3.1), cannot be reconciled with the Kraut–Kennedy relation (3.3), which he thus rejects.

In a later paper, Boschi (1974b) re-estimated the melting curve for iron at high pressures by two other methods. In the first he used a melting equation

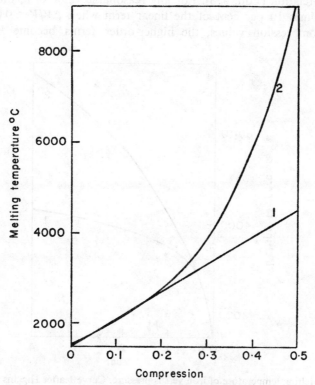

Fig. 3.6. Melting temperature of iron versus $\Delta V/V_0$. Curve 1 after Higgins and Kennedy (1971). Curve 2 after Boschi (1974b).

derived from the Ross–Lindemann melting criterion (Ross, 1969). Ross generalized the Lindemann melting law by reformulating it in terms of the statistical-mechanical partition function: for a given substance, at all points along its melting curve, the solid always occupies the same fraction of configurational phase space. Boschi obtained the equation

$$T_m = T_{m_0} (1 - \Delta V / V_0)^{-n/3}. \tag{3.5}$$

He obtained the same equation independently from the ideal "three phase model" proposed by Hiwatari and Matsuda (1972a, b), using the results of Monte Carlo calculations by Hoover $et\ al.$ (1970). Expanding Eqn. (3.5) gives

$$T_m = T_{m_0} \left\{ 1 + \frac{n}{3} \frac{\Delta V}{V_0} + \frac{n}{6} \left(\frac{n}{3} + 1 \right) \left(\frac{\Delta V}{V_0} \right)^2 + \ldots \right\} \tag{3.6}$$

which, if second- and higher-order terms are neglected, reduces to the Kraut–Kennedy relation (Eqn. 3.3). For any reasonable value of n, the quadratic term becomes 10 per cent of the linear term when $\Delta V / V_0 \approx 0.05$ and for greater compression values, the higher-order terms become increasingly

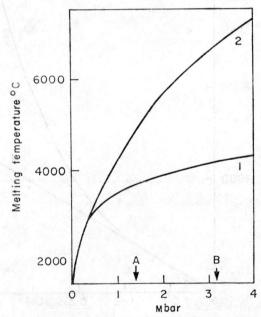

Fig. 3.7. Melting temperature of iron versus pressure. Curve 1 after Higgins and Kennedy (1971). Curve 2 after Boschi (1974b). A and B correspond to pressures at the MCB and inner core–outer core boundaries respectively.

important. Boschi took $n = 8 \cdot 4$, the value obtained by Hiwatari and Matsuda from a study of isothermal compression in the solid phase.

Figure 3.6 shows the melting temperature of iron against compression as obtained by Boschi with the corresponding curve of Higgins and Kennedy included for comparison: Fig. 3.7 gives the corresponding melting temperatures of iron as a function of pressure. Boschi estimated the melting temperature of pure iron to be $\sim 4800°C$ at the MCB and $6600°C$ at the inner core–outer core boundary. The melting point gradient ($\sim 0 \cdot 8°C/km$) across the

Table 3.1. (After Boschi 1974b.)

Method	Iron melting temperature at 1·4 Mbar (°C)	Reference
	3120	
	2950	Simon (1953)
	2810	
Simon equation with different	3320	
values of the empirical	2950	Bullard (1954)
parameters	4610	
	3920	Gilvarry (1956c)
	3360	
	2340	Strong (1959)
	4750	Birch (1963)
	4000	Valle (1955)
Lindemann equation with semi-	4250	Zharkov (1959)
empirical considerations	4400	Alder (1966)
Kraut–Kennedy equation	3750	Higgins and Kennedy (1971)
Significant structure theory	4700	Leppaluoto (1972)
Ross–Lindemann criterion	4800	Boschi (1947b)

outer core is thus considerably steeper than that of Higgins and Kennedy and should well exceed the adiabatic temperature gradient. Table 3.1 gives a list of recent estimates of the melting temperature of iron at MCB conditions. It must not be forgotten, however, that the core is not pure iron, but contains 15 per cent of some light alloying element (probably Si or S, see Section 5.5). This would modify any phase relationships, and also considerably lower melting temperatures in the outer core—Hall and Murthy (1972) suggest that the eutectic temperature may be some 1600°C lower than that for pure iron. There is also evidence that the depth of the eutectic trough deepens with increasing pressure (Kim et al., 1972).

3.4. Adiabatic Temperatures

The increase in temperature dT for a reversible adiabatic increase of pressure dp is given by

$$dT = \frac{T\alpha}{\rho c_p}\, dp \qquad (3.7)$$

where α is the volume coefficient of thermal expansion, ρ the density and c_p the specific heat at constant pressure. Assuming hydrostatic equilibrium, the variation of pressure with depth z is

$$\frac{dp}{dz} = g\rho. \qquad (3.8)$$

Hence the adiabatic temperature gradient is given by

$$\frac{dT}{dz} = \frac{g\alpha T}{c_p}. \qquad (3.9)$$

Values of α and c_p are not well known in the core—the range of values (in kg J^{-1}) as quoted by various authors (see Frazer, 1973) is

$$0\cdot48 \times 10^{-8} \leqslant \frac{\alpha}{c_p} \leqslant 2\cdot2 \times 10^{-8}. \qquad (3.10)$$

Since values of $g = g(z)$ are sufficiently well known, it is possible to integrate Eqn. (3.9) for different values of α/c_p, assuming that the adiabatic and melting temperatures are the same at the inner core–outer core boundary. Adiabatic and melting temperatures can thus be compared throughout the outer core. Assuming Boschi's (1974b) estimates of the melting point of iron at the inner core–outer core boundry (6600°C) and at the MCB (4800°C), α/c_p must be less than $1\cdot8 \times 10^{-8}$ for the adiabatic temperature to exceed the melting temperature at the MCB. If the melting point of iron is lowered because of the addition of lighter alloying elements in the core, the upper bound for α/c_p is increased. It is obvious, however, from Eqn. (3.9) that the adiabatic temperature gradient is critically dependent on the value of α/c_p and its possible radial dependence.
    Writing

$$\Gamma = \frac{\alpha k_s}{\rho c_p} \qquad (3.11)$$

where k_s is the adiabatic incompressibility, the adiabatic temperature gradient (Eqn. 3.9) may be written

$$\frac{dT}{dz} = \frac{gT\rho\Gamma}{k_s} = \frac{gT\Gamma}{\phi} \qquad (3.12)$$

where $\phi = k_s/\rho = V_P^2 - \frac{4}{3}V_S^2$ (see Eqn. 1.3) and is known from seismic data. There have been many estimates of the adiabatic temperature gradient using Eqns. (3.9) or (3.12). The most recent estimate of Higgins and Kennedy (1971) uses Eqn. (3.12) and a value of 1·75 for Γ (Knopoff and Shapiro, 1969). They obtained an alternative estimate based on a relationship derived by Valle (1952).

The adiabatic temperature gradient is quite sensitive to the assumed value of Γ for iron at high pressures—in fact for a liquid, Γ may be no more than a dimensionless combination of thermodynamic parameters and have no real connection with Grüneisen's parameter γ for a solid (Knopoff and Shapiro, 1969). Birch (1972) has also pointed out some of the problems involved in calculating adiabats, e.g. the deflection of an adiabat on passing through a phase boundary and the usual (unverified) assumption that the Grüneisen parameter γ is independent of temperature. Birch conjectured that the γ-phase of iron is suppressed or confined to low pressures and that the ϵ-phase will be the stable solid phase at core pressures—in this case he estimated that the adiabat originating at the melting temperature of the inner core–outer core boundary lies entirely in the liquid phase, though never very far from the freezing curve—supporting the earlier (1971b) conclusions of Jacobs.

Verhoogen (1973) has also estimated the adiabatic temperature gradient from Eqn. (3.12) using a method due to Stewart (1970). Stewart showed that our present knowledge of the density and distribution of ϕ in the outer core is not sufficiently accurate to allow a unique determination of the parameters that characterize core material. He was able to construct a number of graphs showing what combinations of values of the parameters fit the seismic data with a given accuracy. He found that a very wide range of values of Grüneisen's ratio γ are consistent with the seismic data and the assumed density–pressure distribution in the core, and concluded that γ in the core cannot be estimated from seismic data alone or from its value for solid iron at low pressures. Of the acceptable sets of parameters, many yield temperatures at the MCB in the range 3500–5000°K. Some give an adiabatic gradient steeper than the conductive gradient and are thus compatible with convection, others do not. Since the properties of Fe–S melts are unknown it is not possible to select any one set of parameters in preference to another or even to assert that the outer core is a Fe–S melt (see Section 5.5). The important point is however, as Verhoogen pointed out, that solutions can be found which correspond to near adiabatic or super adiabatic conditions in the outer core which are therefore compatible with convection.

Stacey (1972) suggested that if the temperature gradient of the liquid in the outer core of the Earth is adiabatic, heat flowing down the adiabat would have to escape into the mantle and that, unless there are special heat sources in the core and special heat sinks in the mantle, the core of the Earth would soon

become isothermal. Kennedy and Higgins (1973b) have come essentially to
the same conclusion, estimating a temperature difference of only 25–35°C
across the outer core. This would completely inhibit any kind of adiabatic
radial mixing of the core.

A way out of this conclusion would be if there were a heat source in the
Earth's core and a heat sink at the MCB. Freezing of iron at the inner core–

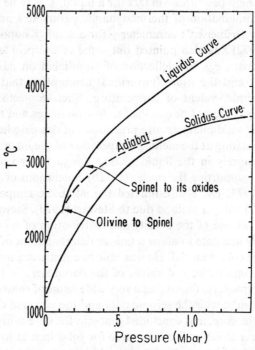

Fig. 3.8. Estimated solidus, liquidus and adiabat for the Earth's mantle. (After Kennedy
and Higgins, 1973b.)

outer core boundary (Verhoogen, 1961) and/or potassium in the fluid outer
core (Murthy and Hall, 1972) might possibly provide sufficient heat sources.
Convective circulation of deep mantle rocks might carry away sufficient heat
to enable a high temperature gradient to be maintained in the core. An
alternative heat sink in the deep mantle was proposed by Kennedy and
Higgins (1973b) who re-estimated the liquidus, solidus and adiabatic gradient
in the Earth's mantle (see Fig. 3·8). They found that the adiabatic and solidus
curves are essentially the same over the lower one-third of the mantle,
suggesting that convection is restricted to the upper two-thirds. Also the

temperature of their solidus curve is not too different at the MCB from the estimated temperature of the melting of iron. They thus suggested that the requisite heat sink could be formed by the melting of silicate or oxide materials near the MCB with upward migration of liquid along the melting curve. This would imply that the temperature gradient throughout the mantle is that of the solidus curve. Seismological evidence gives some support to the possibility of melting at the MCB—a number of authors have reported a substantial decrease in shear velocity immediately above the MCB (see Section 1.2).

3.5. The Earth's Inner Core Reconsidered

Higgins and Kennedy (1971) found that both the melting-point and adiabatic gradients were extremely flat in the inner core (see Fig. 3.4). Assuming that the inner core is solid and the outer core fluid, so that the adiabatic and melting temperatures are the same at the inner core–outer core boundary, Higgins and Kennedy found that these temperatures differ by only 15°C at the centre of the Earth. On the other hand they found a very sharp curvature in the adiabatic gradient in the outer core—much steeper than the melting-point gradient—the adiabatic gradient being about 1250°C across the outer core, compared with 500°C for the melting-point gradient. This is just the opposite to what has usually been supposed, and the mechanism which Jacobs (1953a) put forward for the formation of a solid inner core and liquid outer core would no longer be valid. If actual temperatures were distributed along the adiabat of Higgins and Kennedy throughout the outer core, it too would be solid—there would be no liquid outer core. Higgins and Kennedy thus concluded that the actual temperature gradient in the outer core is much less than the adiabatic gradient. If this is the case the outer core would be thermally stably stratified, thereby inhibiting radial convection and the question of the generation of the motions which drive the geomagnetic dynamo would have to be reconsidered (see Section 4.7).

The immediate reaction to Kennedy and Higgins' paper was to accept their results at face value and try to invent ways and means to get around what they have since called the core paradox (1973a). Thus Bullard and Gubbins (1971) pointed out that a stable fluid can have internal wave motions, and they carried out preliminary calculations which indicated that a body of fluid could possibly act as a dynamo, even when its motion is purely oscillatory. Busse (1972), Malkus (1973) and Elsasser (1972) independently suggested that the outer core might consist of a slurry of fine iron particles suspended in an iron-rich liquid. There is however a critical limit to the solid grain size in order that they do not precipitate out faster than the core can stir them up. Malkus (1973) estimated that the critical size is 1 micron for a convection-driven dynamo and 10 microns for a precession-driven dynamo. Such small grain sizes do not seem to be in accord with metallurgical experience. Reports of extremely low attenuation of seismic waves in the outer core also seem hard to reconcile with such a constitution.

Stacey (1972) has suggested yet another solution. If the light alloying element in the core is sulphur (see Section 5.5) and if this is confined to the outer core, the inner core consisting mainly of nickel–iron with very few lighter components, the presence of sulphur in the outer core may so reduce its liquidus below that of iron that the adiabat of Higgins and Kennedy through the inner core–outer core boundary does not intersect it (see Fig. 3.9).

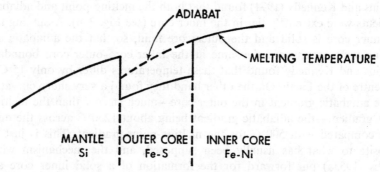

Fig. 3.9. Possible melting-point curve in the Earth and adiabat through the inner core–outer core boundary. (After Stacey, 1972.)

Some of the difficulties in estimating the melting-point and adiabatic temperatures in the core have already been discussed in Section 3.3 and 3.4. Jacobs (1971b) has re-estimated the adiabatic gradient in the core by a different method (1953b)—assuming, by analogy with Bullen's (1946) compressibility—pressure hypothesis, that there is a linear relationship between $1/\alpha$, where α is the coefficient of thermal expansion, and pressure. There is no real reason to suppose that the use of such an empirical relationship is superior to any of the other methods. However Jacobs' calculations indicate that in the core the adiabatic temperature gradient is less than the melting-point gradient of Higgins and Kennedy (see Fig. 3.4). More detailed calculations by Birch (1972) have led to the same conclusion viz. that "the adiabat originating at the melting temperature of the inner core–outer core boundary lies entirely in the liquid phase". In view of all the uncertainties in the estimates of both the melting-point and adiabatic temperatures, it is impossible to say definitely which gradient is the steepest in the core (Jacobs, 1973). It thus seems unnecessary at the moment to try and circumvent the consequences of accepting Higgins and Kennedy's results.

Both Jacobs (1971a) and Birch (1972) have concluded that actual temperatures in the core are probably very close to those of the melting temperature. If the melting-point and adiabatic gradients are virtually the same throughout the entire core, then perhaps parts of the mantle may from time to time become soluble in the outer core or material from the outer core diffuse into the lower mantle—thereby giving rise to "bumps" on the core–mantle boundary as has been suggested by Hide (1969) in a different connection (Section 4.10). The position and shape of the boundary may thus change over the course of time and be instrumental in initiating and dictating motions in the outer core. If the topography of either boundary of the outer core has a direct influence on core motions, then changes in the frequency of reversals of the Earth's magnetic field may well be random as the shape of these boundaries (randomly) changes. It is not easy to estimate the time scale for producing such bumps on the core boundary, but it seems likely that it is greater than the (geologically) short time scale of reversals found for the last 20 m yr (Section 4.5). If this is the case, then there could not be a one-to-one correspondence between the production of bumps and individual reversals. However changes in the topography of the core boundary could indirectly affect the frequency of reversals of the Earth's magnetic field—perhaps accounting for such intervals in the Earth's history as the Kiaman (a span of some 50–60 m yr, some 300 m yr ago) when reversals have only very rarely been observed.

Schloessin (1974) has put forward the interesting suggestion that bumps on the MCB may arise from constitutional supercooling. Constitutional supercooling is generally found in crystallization from impure solutions or melts. It is favoured by constraint growth conditions, low temperature gradients across the solid–liquid interface and high viscosity of the liquid phase. Rejection of an impurity component from the solid interface raises the impurity content in the liquid near the interface. If, owing to the phase relationship of the components involved, the impurity causes an increased liquidus temperature, the actual temperature in the liquid, ahead of the interface, will be below the liquidus temperature and the liquid becomes "constitutionally supercooled". As a result the solid–liquid interface becomes unstable and advances. Schloessin suggested that there may be a slow growth of the mantle and inner core at the expense of the liquid outer core by such a process. Although formation of a solid inner core and liquid outer core most probably occurred early in the Earth's history (see Section 2.4), the liquid phase may have remained sufficiently impure for a slow "overgrowth" of both mantle and inner core to be still continuing. The abnormal P velocities in both the D'' and F layers (see Section 1.2) may be evidence for a slow growth rate of both boundaries of the outer core.

The advance of the solid–liquid interface will be controlled by T–T_L where T is the actual temperature and T_L the liquidus temperature. If T_L is sensitive to the concentration C_{Li} of an impurity component i in the liquid and $dT_L/dC_{Li} > 0$, then regions rich in component i will solidify at a lower temperature and regions depleted in i, at a higher temperature. For a given radial variation of C_{Li}, the position of the liquidus will follow roughly the inverse of the curve for the impurity distribution (see Fig. 3.10). Over a

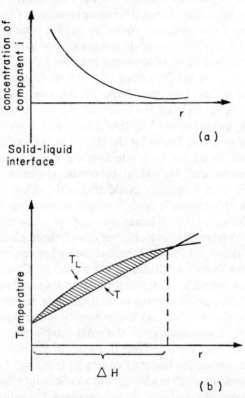

Fig. 3.10. Variation of the concentration of a rejected component (i) and of the corresponding liquidus temperature (T_L) as a function of distance (r) from the solid–liquid interface. The shaded region, extending to a distance ΔH, determines the width of the constitutionally supercooled region. (After Schloessin, 1974.)

distance ΔH to the right of the interface, $T_L > T$, thus defining the region which is constitutionally supercooled.

Schloessin also suggested that constitutional supercooling may provide the motive force for driving the geodynamo (Section 4.7) through the free energy change associated with the overgrowth of both the mantle and the inner core. Density inhomogeneities could be generated by this mechanism independent of thermal inhomogeneities—even in a predominantly stratified core. Self sustained and regenerative concentration gradients of segregated components confined either to layers adjacent to the boundaries or extending through the entire liquid core could thus be the answer to the core paradox of Kennedy and Higgins (1973a).

3.6. The Core and the Thermal History of the Earth

The classical approach to the Earth's thermal history is to formulate it as an initial boundary value problem with calculations based on the theory of heat conduction in a solid. Although calculations based on the theory of heat conduction are relatively straightforward, the results are not applicable to the real Earth. The data required are poorly known and conduction does not describe all the processes of heat and mass transfer within the Earth. Large scale convection is extremely efficient in transporting heat and such convective heat transfer will dominate thermal lattice and radiative heat conduction even for a small velocity, of the order of 10^{-2} cm/yr. Both MacDonald (1963) and Knopoff (1964) have presented evidence against large scale convection in the mantle, but since their arguments depend quite critically on the assumed rheological behaviour of the Earth, they cannot be regarded as conclusive.

An investigation of the thermal history of the Earth taking into account convection and fractionation of radioactive elements is extremely difficult. The difficulties are two-fold—mathematical and physical. A mathematical treatment of the problem entails the formulation and solution of the complete field equations of a multi-component, multi-phase and radioactive continuum of varying properties. The physical difficulties arise mainly from our lack of understanding of the Earth's rheological behaviour and fractionation processes. The most detailed discussion of this problem is due to Lee (1967, 1968) who developed mathematical techniques to treat the Earth's thermal history beyond simple heat-conduction theory, taking into account latent heat, convection, and fractionation of radioactive elements. Lee showed that large scale convection within the Earth is unlikely and that heat transfer by small scale penetrative convection is also unimportant. Such convection is of great importance, however, as a means of moving the radiogenic heat sources upward.

Very little work has been done on convection as a mode of heat transfer in the mantle although it has been realized that convection may play a dominant role in the thermal regime in the Earth. Tozer (1967, 1970a, b) has made some significant contributions in this regard, although the value of his work has unfortunately not been fully recognized or appreciated.

As a result of both experimental work and theoretical considerations, Tozer proposed that convective motions in the Earth are such as to minimize the mean temperature across any spherical shell in which the conduction solution is unstable and that it is possible to use data from laboratory model experiments to find that minimum. Tozer also showed that it is possible to estimate mean temperatures on level surfaces without knowing the exact details of the velocity distribution—such temperatures below a depth of about 800 km are controlled by the viscosity dependence on temperature. For all plausible viscosity–temperature relationships, the temperature is always that which gives a viscosity $\simeq 10^{20}$–10^{21} poise. Thus the prevalent idea that any convection theory for the mantle must be very imprecise because of uncertainty in the viscosity–temperature relationship is not true—rather the viscosity itself is constrained to lie within very narrow limits.

In a later paper, Tozer (1972) showed that for a very large range of physically plausible values of the material parameters, the temperature distribution in bodies larger than about 800 km in radius is very different from that predicted by conduction theory. In any body the average temperature rises with depth according to the conduction or state of rest solution until either the centre of the body is reached or the kinematic viscosity has fallen to a value $\simeq 10^{20}$ cm^2/sec—whichever is reached first. Once a body is large enough

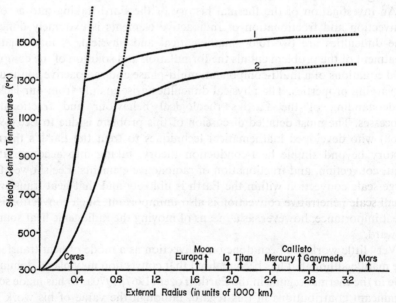

Fig. 3.11. The central temperature T_c as a function of the external radius R for two values of the heat source density H. Curve 1: $H = 1 \cdot 6 \times 10^{-14}$ cal/cm^3 sec, Curve 2: $H = 5 \times 10^{-15}$ cal/cm^3 sec. On the left the steeply rising curves are steady state of rest solutions—unstable where dotted. Note the relative independence of T_c on R when $R > 800$ km and the surprisingly low values of T_c. (After Tozer, 1972.)

for its central viscosity to be incapable of stabilizing a state of rest solution, the steady central temperature becomes comparatively independent of the radius and surprisingly low (see Fig. 3.11). Thus of all bodies in the solar system, small objects (radius $\lesssim 800$ km) best preserve conditions existing at their birth. The magnitude of the velocity of internal motions, the non-hydrostatic stresses and the viscosity in the convective zones are all relatively insensitive to the choice of values of the material parameters.

A difficult feature of any calculation of the thermal regime of the Earth based on conduction theory alone is that the thermal time constant is much greater than the age of the Earth with the result that temperatures in the deep interior depend on the thermal conditions existing at the time of formation of the Earth. This difficulty is largely overcome by convection theory. A stabilization temperature of less than about 2500°K and the great increase in heat transport above it, result in the present thermal conditions being virtually independent of initial conditions, and make the temperature distribution far less sensitive to details of the heat source distribution.

Verhoogen (1973) has shown that it is possible to construct models of a radioactive core (due to the decay of K^{40}) which agree with seismic data and account for convective instability, the existence of a solid inner core and other details of core structure (particularly the F layer). If the average magnetic field in the core is 100 gauss and has a characteristic decay time of 10^4 yr, (see Section 4.7), maintenance of that field requires an energy supply at a rate of 2×10^{17} erg/sec. The efficiency of the geodynamo is not known but may well be less than 1 per cent. Verhoogen assumed that the outer core consists of Fe, S and a small amount of K (see Section 5.5) which generates heat by the radioactive decay of K^{40}. He considered two cases, one corresponding to a high rate of heat production h (8×10^{19} erg/sec $\approx 2 \times 10^{12}$ cal/sec) which would require about 0·1 per cent K by weight. This value of h corresponds to a heat flux at the surface of the Earth of about 25 per cent of the average measured value. Verhoogen took as the other case a lower limit for h (10 times smaller, corresponding to a rate of heat production of 2×10^{11} cal/sec).

The temperature in the lower mantle has been estimated by Wang (1972) to be about $3300°K \pm 800°$ at a depth of 2800 km. Bolt (1972) has shown that the decrease with depth of the seismic velocities V_P and V_S in the lowermost 100–150 km of the mantle (region D'') may imply a sharp density gradient. This density gradient, caused perhaps by an admixture of core material may be sufficiently steep to prevent convection and maintain a high (conductive) temperature gradient which accounts in part for the decrease in the seismic velocities. The layer D'' may then be thought of as a non-convecting thermal boundary layer separating the convecting thermal regime of the core from that of the lower mantle above D''. If this interpretation is correct the temperature T_c at the MCB may be some 1300–1900° higher than the temperature at the top of D″. Verhoogen thus took T_c to be about 4500–5000°K for $h = 2 \times 10^{12}$ cal/sec and perhaps only 3500°K for the lower limit of h or if the layer D″ participates in mantle convection.

Instability in the outer core probably requires a temperature gradient at least as great as the adiabatic. Verhoogen thus first determined the gradient that would be maintained by the heat sources in the absence of convection. For distributed heat sources (K^{40}) the temperature difference across the outer core is approximately 1600°K for high h and only 160°K for low h. With this lower value of h the core is essentially isothermal. The heat drop across the core could be raised to 850° even for this low value of h if it was released by crystallization at the inner core–outer core boundary (Verhoogen, 1961), rather than by the radioactive decay of K^{40}. Verhoogen estimated the adiabatic gradient from Eqn. (3.9) and showed that the lower rate of heat production from distributed heat sources leads to a conductive gradient everywhere less than the adiabatic and is thus incompatible with convection.

Verhoogen also pointed out that the composition of the liquid that could be in equilibrium with solid iron at the inner core boundary is further complicated by the likely occurrence of liquid immiscibility. Kullerud (1970) has pointed out that most sulphur–metal systems exhibit liquid immiscibility, i.e. the possible co-existence of two liquids of different composition. The likelihood of a miscibility gap in Fe–S liquids of low S content at high pressures and at temperatures close to the liquidus may help in explaining peculiarities in core structure in the F region near the inner core boundary where Jeffreys and Bullen suggested that the compressional wave velocity decreased with depth (see Section 1.2). The properties of this region appear to differ only slightly from those in region E, suggesting that F is also liquid.

Sacks (1972) found that high values (about 5000) of the Q factor that characterize region E (see also Section 1.2) persist in F down to the inner core boundary as would probably not be the case if F were a two-phase solid–liquid mush. On the other hand if F consists of a liquid with properties different from those of region E and is separated from E by a relatively sharp boundary, liquid immiscibility is implied. Verhoogen suggested that F consists of an Fe rich liquid F in equilibrium with both solid Fe and another liquid E richer in S than F. In this interpretation region F is poor in S and rich in Fe. The temperature T_i at the inner core–outer core boundary cannot then be much lower than the melting-point of the inner-core material. The same conclusion holds if, as suggested by Usselman (1972), the eutectic composition is much richer in Fe at high pressures than at low pressures.

Verhoogen pointed out that there are other signs that T_i is close to the melting point of inner-core material, assumed to be mostly Fe with a few per cent Ni. There is the low value of Q (~ 600) found by Sacks (1972) in the outermost part of the inner core compared with the higher value (~ 3000) at the centre of the Earth. There is also the low value of the rigidity μ of the inner core compared with its bulk modulus k_s and correspondingly high Poisson's ratio σ (0·45). Finally there is a rapid increase with depth in the inner core of k_s (Bolt, 1972). These observations are consistent with the rapid drop in μ, k_s and Q and the rise in σ observed in metals just below their melting-point (Mizutani and Kanamori, 1964).

Verhoogen's final conclusion is that the data indicate that the central region G consists of solid Fe (or Ni–Fe) in equilibrium with a liquid F containing perhaps as little as 2–3 per cent S, while the liquid region E contains more S than the region F, perhaps as much as 10 per cent. It thus appears possible to construct satisfactory models of the core if it is assumed to consist of Fe with about 10 per cent S and about 0·1 per cent K in the outer core. (The composition of the core will be discussed in more detail in Chapter 5.) Such models are in good agreement with data on the density and seismic velocity distributions, heat flow and rate of production of magnetic energy. They predict temperatures of the order of 3500–5000°K at the MCB and of 5000–6500°K at the inner core–outer core boundary. In some of the models the temperature gradient in the outer core slightly exceeds the adiabatic so that convection can occur if the inhibiting effects of the Earth's rotation and magnetic field are not significant.

References

Alder, B. J. (1966). Is the mantle soluble in the core? *J. geophys. Res.* **71**, 4973.

Birch, F. (1963). Some geophysical applications of high pressure research. *In* "Solids Under Pressure" (Eds. W. Paul and D. M. Warschauer), McGraw-Hill, New York.

Birch, F. (1972). The melting relations of iron and temperatures in the Earth's core. *Geophys. J.* **29**, 373.

Bolt, B. A. (1972). The density distributions near the base of the mantle and near the Earth's centre. *Phys. Earth Planet. Int.* **5**, 301.

Boschi, E. (1974a). On the melting curve at high pressures. *Geophys. J.* **37**, 45.

Boschi, E. (1974b). Melting of iron. *Geophys. J.* **38**, 327.

Bullard, E. C. (1954). The interior of the Earth. *In* "The Earth as a Planet", University of Chicago Press, Illinois.

Bullard, E. C. and Gubbins, D. (1971). Geomagnetic dynamos in a stable core. *Nature* **232**, 548.

Bullen, K. E. (1946). A hypothesis on compressibility at pressures of the order of a million atmospheres. *Nature* **157**, 405.

Busse, F. H. (1972). Comment on the paper "The adiabatic gradient and the melting point gradient in the core of the Earth" by G. Higgins and G. C. Kennedy. *J. geophys. Res.* **77**, 1589.

Elsasser, W. M. (1972). Thermal stratification and core convection, Int. Conf. Core–Mantle Interface, *Trans. Am. Geophys. Union* **53**, 605.

Frazer, M. C. (1973). Temperature gradients and the convective velocity in the Earth's core. *Geophys. J.* **34**, 193.

Gilvarry, J. J. (1956a). The Lindemann and Grüneisen laws. *Phys. Rev.* **102**, 308.

Gilvarry, J. J. (1956b). Grüneisen's law and the fusion curve at high pressures. *Phys. Rev.* **102**, 317.

Gilvarry, J. J. (1956c). Equation of the fusion curve. *Phys. Rev.* **102**, 325.

Gilvarry, J. J. (1957). Temperatures in the Earth's interior. *J. atmos. terr. Phys.* **10**, 84.

Gilvarry, J. J. (1966). Lindemann and Grüneisen laws and a melting law at high pressure. *Phys. Rev. Letters* **16**, 1089.

Hall, H. T. and Murthy, V. R. (1972). Comments on the chemical structure of an Fe–Ni–S core of the Earth, Int. Conf. Core–Mantle Interface, *Trans. Am. Geophys. Union* **53**, 602.

Hide, R. (1969). Interaction between the Earth's liquid core and solid mantle, *Nature* **222**, 1055.

Higgins, G. H. and Kennedy, G. C. (1971). The adiabatic gradient and the melting point gradient in the core of the Earth. *J. geophys. Res.* **76**, 1870.

Hiwatari, Y. and Matsuda, H. (1972a). Ideal three-phase model and the melting of molecular crystals and metals. *Prog. theor. Phys.* **47**, 741.

Hiwatari, Y. and Matsuda, H. (1972b). "Ideal three-phase model" and the melting of molecular crystals and metals. *In* "The Properties of Liquid Metals" (Ed. S. Takeuchi), Taylor and Francis, London.

Hoover, W. G., Ross, M., Johnson, K. W., Henderson, D., Barker, J. A. and Brown, B. C. (1970). Soft sphere equation of state. *J. chem. Phys.* **52**, 4931.

Jacobs, J. A. (1953a). The Earth's inner core. *Nature* **172**, 297.

Jacobs, J. A. (1953b). Temperature-pressure hypothesis and the Earth's interior. *Can. J. Phys.* **31**, 370.

Jacobs, J. A. (1971a). Boundaries of the Earth's core. *Nature Phys. Sci.* **231**, 170.

Jacobs, J. A. (1971b). The thermal regime in the Earth's core. *Comments Earth Sci. Geophys.* **2**, 61.

Jacobs, J. A. (1973). Physical state of the Earth's core. *Nature Phys. Sci.* **243**, 113.

Jayaraman, A., Newton, R. C. and McDonough, J. M. (1967). Phase relations, resistivity and electronic structure of cesium at high pressures. *Phys. Rev.* **159**, 527.

John, M. S. and Eyring, H. (1971). The significant structure theory of liquids. *In* "Physical Chemistry, An Advanced Treatise", Vol. VIIIa (Ed. D. Henderson), Academic Press, New York.

Kennedy, G. C. and Vaidya, S. N. (1970). The effect of pressure on the melting points of solids. *J. geophys. Res.* **75**, 1019.

Kennedy, G. C. and Higgins, G. H. (1973a). The core paradox. *J. geophys. Res.* **78**, 900.

Kennedy, G. C. and Higgins, G. H. (1973b). Temperature gradients at the core–mantle interface. *The Moon* **7**, 14.

Kim, Ki-Tae, Vaidya, S. N. and Kennedy, G. C. (1972). Effect of pressure on the temperature of the eutectic minimums in two binary systems: NaF–NaCl and CsCl–NaCl. *J. geophys. Res.* **77**, 6984.

Knopoff, L. (1964). The convection current hypothesis. *Rev. Geophys.* **2**, 89.

Knopoff, L. and Shapiro, J. N. (1969). Comments on the inter-relationship between Grüneisen's parameter and shock and isothermal equations of state. *J. geophys. Res.* **7**, 1439.

Kraut, E. A. and Kennedy, G. C. (1966). New melting law at high pressures. *Phys. Rev.* **151**, 668.

Kullerud, G. (1970). Sulphide phase relations. *Min. Soc. Am. Spec. Publ.* **3**, 199

Lee, W. H. K. (1967). The thermal history of the Earth. Ph.D. Thesis, University of California, Los Angeles.

Lee, W. H. K. (1968). Effects of selective fusion in the thermal history of the Earth's mantle. *Earth Planet. Sci. Letters* **4**, 270.

Leppaluoto, D. A. (1972). Melting of iron by significant structure theory. *Phys. Earth Planet. Int.* **6**, 175.

MacDonald, G. J. F. (1963). The deep structure of the continents. *Rev. Geophys.* **1**, 587.

Malkus, W. V. R. (1973). Convection at the melting point, a thermal history of the Earth's core. *Geophys. Fluid Dyn.* **4**, 267.

McLachlan, D. and Ehlers, E. G. (1971). Effect of pressure on the melting temperature of metals. *J. geophys. Res.* **76**, 2780.

Mizutani, H. and Kanamori, H. (1964). Variations of elastic wave velocity and attenuation properties near the melting point. *J. Phys. Earth (Tokyo)* **12**, 43.

Murthy, V. Rama and Hall, H. T. (1972). The origin and chemical composition of the Earth's core. *Phys. Earth Planet. Int.* **6**, 123.

Ree, F.H. (1971). Computer calculations for model systems. *In* "Physical Chemistry, An Advanced Treatise", Vol. VIIIa (Ed. D. Henderson), Academic Press, New York.

Ross, M. (1969). Generalized Lindemann melting law. *Phys. Rev.* **184**, 233.

Sacks, I. S. (1972). *Q*-structure of the inner and outer core, Int. Conf. Core–Mantle Interface, *Trans. Am. Geophys. Union* **53**, 601.

Salter, L. (1954). The Simon melting equation. *Phil. Mag.* **45**, 369.

Schloessin, H. H. (1974). Corrugations on the core boundary interfaces due to constitutional supercooling and effects on motion in a predominantly stratified liquid core. *Phys. Earth Planet. Int.* **9**, 147.

Simon, F. E. (1937). On the range of stability of the fluid state, Trans. Faraday Soc. **33**, 65.

Simon, F. E. (1953). The melting of iron at high pressures. *Nature* **172**, 746.

Stacey, F. D. (1972). Physical properties of the Earth's core. *Geophys. Surv.* **1**, 99.

Sterrett, K. F., Klement, W., Jr. and Kennedy, G. C. (1965). The effect of pressure on the melting of iron. *J. geophys. Res.* **70**, 1979.

Stewart, R. M. (1970). Shock wave compression and the Earth's core. Ph.D. thesis, University of California, Berkeley.

Strong, H. M. (1959). The experimental fusion curve of iron to 96000 atmospheres. *J. geophys. Res.* **64**, 653.

Tozer, D. C. (1967). Towards a theory of thermal convection in the mantle. *In* "The Earth's Mantle" (Ed. T. F. Gaskell), Academic Press, London.

Tozer, D. C. (1970a). Factors determining the temperature evolution of thermally convecting Earth models. *Phys. Earth Planet. Int.* **2**, 393.

Tozer, D. C. (1970b). Temperature, conductivity, composition and heat flow. *J. Geomagn. Geoelect.* **22**, 35.

Tozer, D. C. (1972). The present thermal state of the terrestrial planets. *Phys. Earth Planet. Int.* **6**, 182.

Tuerpe, D. R. and Keeler, R. N. (1967). Anomalous melting transition in the significant structure theory of liquids. *J. chem. Phys.* **47**, 4283.

Usselman, T. M. (1972). The Fe–FeS system at high pressures and the chemical zonation of the core, Int. Conf. Core–Mantle Interface, *Trans. Am. Geophys. Un.* **53**, 603.

Valle, P. E. (1952). Sul gradiente adiabatico di temperatura nell-interno della terra. *Annali Geofis.* **5**, 41.

Valle, P. E. (1955). Una stima del punto di fusione del ferro sotto alte pressioni. *Annali Geofis.* **8**, 189.

Van Thiel, M. (ed.) (1966). Compendium of shock wave data, University of California, Livermore 50108.

Verhoogen, J. (1961). Heat balance of the Earth's core. *Geophys. J.* **4**, 276.

Verhoogen, J. (1973). Thermal regime of the Earth's core. *Phys. Earth Planet. Int.* **7**, 47.

Wang, C.-Y. (1972). Temperatures in the lower mantle. *Geophys. J.* **27**, 29.

Zharkov, V. N. (1954). The fusion temperature of the Earth's mantle and the fusion temperature of iron under high pressures. *Bull. Acad. Sci. USSR Geophys. Ser.* **3**, 315.

4. The Earth's Magnetic Field

4.1. Introduction

At its strongest near the poles the Earth's magnetic field is several hundred times weaker than that between the poles of a toy horseshoe magnet—being less than a gauss (Γ). Thus in geomagnetism we are measuring extremely small magnetic fields and a more convenient unit is the gamma (γ), defined as $10^{-5}\Gamma$. Strictly speaking the unit of magnetic field strength is the oersted, the gauss being the unit of magnetic induction. The distinction is somewhat pedantic in geophysical applications since the permeability of air is virtually unity in cgs units. The traditional unit used in geomagnetism, the gauss, has been retained in this book in order to facilitate comparison with values given in the literature. The conversion factor which must be used to convert flux densities (measured in gauss) to SI units is

$$1 \text{ gauss } (\Gamma) = 10^{-4} \text{ weber/m}^2 \text{ (Wb/m}^2)$$

$$= 10^{-4} \text{ tesla (T)}$$

The magnetic field at any point on the Earth's surface may be specified by three parameters, e.g. the total intensity F, declination D and inclination I or the two horizontal components X and Y and the vertical component Z. Simple relationships exist between these different magnetic elements (see e.g. Jacobs, 1963). The variation of the magnetic field over the Earth's surface is best illustrated by isomagnetic charts, i.e. maps on which lines are drawn through points at which a given magnetic element has the same value. Contours of equal intensity in any of the elements X, Y, Z, H or F are called isodynamics. Figures 4.1 and 4.2 are world maps showing contours of equal

115

declination (isogonics) and equal inclination (isoclinics) for the year 1965. It is remarkable that a phenomenon (the Earth's magnetic field) whose origin, as we shall see later, lies within the Earth should show so little relation to the broad features of geology and geography.

DECLINATION (degrees)   D

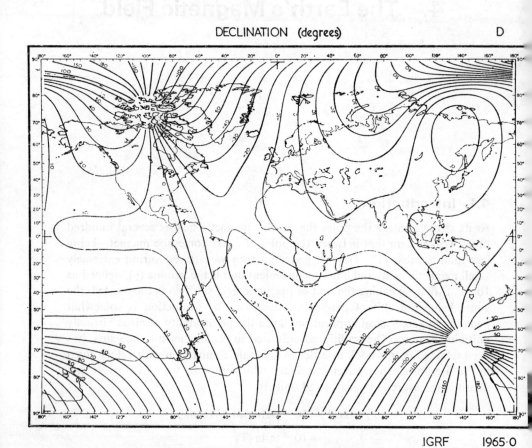

IGRF   1965·0

Fig. 4.1. World map showing contours of equal declination for 1965, International Geomagnetic Reference Field (IAGA Bull. No. 28, 1971)

Not only do the intensity and direction of magnetization vary from place to place across the Earth, but they also show a time variation. There are two distinct types of temporal changes: transient fluctuations and long term secular changes. Transient variations produce no large or enduring changes in the Earth's field and arise from causes outside the Earth. Secular changes, on the other hand are due to causes within the Earth and over a long period of time the net effect may be considerable. If successive annual mean values of a magnetic element are obtained from a particular station, it is found that over a long period of time the changes are in the same sense, although the rate of change does not usually remain constant. Figure 4.3 shows the change in

INCLINATION (degrees)   I

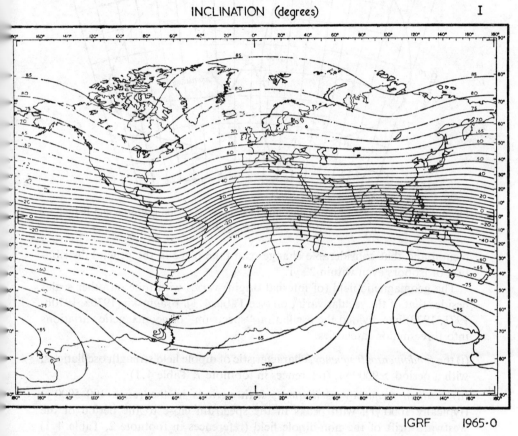

IGRF   1965·0

Fig. 4.2. World map showing contours of equal inclination for 1965, International Geomagnetic Reference Field (IAGA Bull. No. 28, 1971)

declination and inclination at London, Boston and Baltimore. The declination at London was $11\frac{1}{2}°$E in 1580 and $24\frac{1}{4}°$W in 1819, a change of almost 36° in 240 yr. Lines of equal secular change (isopors) in an element form sets of ovals centring on points of local maximum change (isoporic foci). Figures 4.4 and 4.5 show the secular change in Z for the years 1922.5 and 1942.5. It can be seen that the secular variation is a regional rather than a planetary pheno-

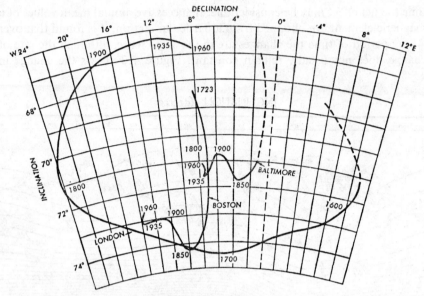

Fig. 4.3. Secular change of declination and inclination at London, Boston and Baltimore. (After Nelson *et al.*, 1962.)

menon and that considerable changes can take place in the general distribution of isopors even within 20 yr.

The geomagnetic field (of internal origin) varies with time on other scales besides that of the secular variation (see Table 4.1). Braginskii (1970a, 1970b, 1971, 1972) has divided the oscillation "spectrum" of geomagnetic variations into three major categories.

(i) the *fundamental frequency*, characteristic of dipole field strength oscillations, with a period $\simeq 9000$ yr. (references in footnote 7, Table 4.1)

(ii) *medium frequency oscillations*, with periods in the range 100–5000 yr (typically 1000 yr), with peaks in the spectrum close to the period of the westward drift of the non-dipole field (references in footnote 2, Table 4.1)

(iii) high frequency oscillations, with periods < 100 yr (references in footnote 1, Table 4.1).

The theoretical treatment of these oscillations is highly complicated. Braginskii suggested that the existence of a fundamental frequency is a consequence of the two-stage nature of the dynamo process, in which a weak poloidal* field leads to the generation of a strong toroidal field, the toroidal field being responsible for the regeneration of the poloidal field. Medium

* For the definition and a discussion of poloidal and toroidal vector fields see Appendix A.

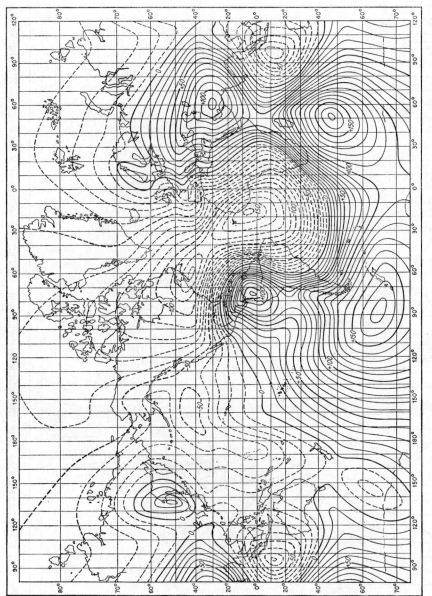

Fig. 4.4. World map showing the geomagnetic secular variation of the vertical component Z. Epoch 1922.5. (After Vestine *et al.*, 1947.)

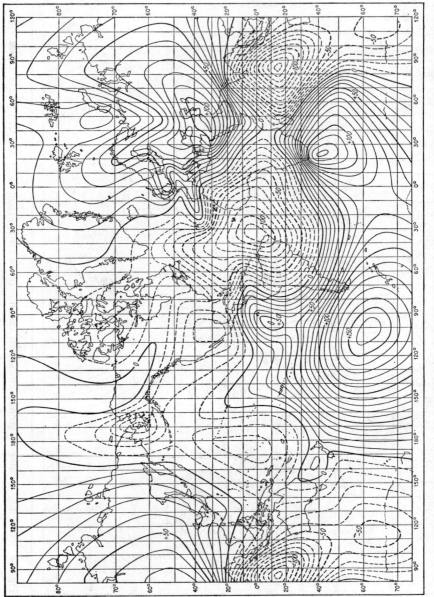

Fig. 4.5. World map showing the geomagnetic secular variation of the vertical component Z. Epoch 1942.5. (After Vestine *et al.*, 1947.)

frequency oscillations are associated with the so-called "MAC-waves" (i.e. Magnetic–Archimedean–Coriolis waves) in the Earth's fluid core (references in footnote 2, Table 4.1). High frequency oscillations are linked to torsional magnetohydrodynamic oscillations and turbulent pulsations in the core (references in footnote 1, Table 4.1).

Currie (1973a) has used the maximum entropy method* to obtain greatly improved knowledge of the geomagnetic spectrum in the period range 2–70 yr. His work provides the first successful spectrum detection of the solar cycle (SC) and double solar cycle (DSC) variations. The structure consists of band spectra centred at 10·5 and 21·4 yr. In addition he was also able to identify the first four harmonics of the SC and, except for one, the first nine harmonics of the DSC. Currie had earlier (1966) shown that the geomagnetic continuum from 40 days to at least 3·7 yr is coherent and in phase over the entire Earth and deduced that it was generated by physical processes external to the Earth. The more detailed analysis of his later paper indicates that fluctuations over two years consist of a number of band spectra harmonics extending all the way to the SC which are a result of SC and DSC sunspot variations on the sun's surface. Currie concluded however that one cannot dismiss the possibility that some of the spectral lines are due to internal signals generated in the Earth's core. In particular the band at 6·07 yr in Fig. 4.6 cannot be ascribed to a harmonic of either fundamental bands and thus may be evidence for internal signals of this period. In this respect it is interesting to note the tendency for secular variation impulses to recur approximately every 5 yr. Yukutake (1972) has also shown that one of the free modes of the electrically coupled core-mantle Earth system has an oscillation period of ∼6·7 yr.

Currie also obtained convincing evidence for a ∼60 yr line in the geomagnetic spectrum. The estimated amplitudes, although rather crude and variable, are too large to be associated with the sun and thus this "line" must represent a signal from the core. In a later paper (1973c), Currie showed that the line is global in character; however the amplitudes of the signal are anomalously low in and around the Pacific basin region. This is also true of the secular variation (Cox and Doell, 1964; Doell and Cox 1971, 1972). A physical explanation for these lows is not clear. The existence of the Pacific quiet (magnetically) zone suggests that generation of irregularities in the core dynamo may be partly controlled by lateral inhomogeneities at the MCB. Such inhomogeneities might be topographic "bumps" at the base of the mantle (see Section 4.10) and could be related to the non-drifting components of the non-dipole field (Yukutake and Tachinaka, 1969; Yukutake, 1970). Evidence

---

* The maximum entropy method is a radically new approach to power spectrum estimation which obviates several limitations of conventional methods and is especially suitable for short time series. It was originally due to Burg (1967, 1968): an outline of the method has been given by Lacoss (1971) and Ulrych (1972).

E

Table 4.1. Geomagnetic field of internal origin: temporal variations. (After Gilliland 1973.)

Type of variation	Time scale (years)	Comments
Secular Variation of Nondipole Field		
High frequency oscillations (1)	<100	Complicated spectrum of variations
Medium frequency oscillations (2)	100–5000	
Westward drift (3)	~2000	
Changes in Energy Distribution		
Dipole–nondipole energy transfer (4)	~2000	In the last 120 yr the net energy loss rate has been ~0.02 per cent/yr, and the dipole–nondipole transfer rate ~0.06 per cent/yr (per cent of total field). However, over the last 1500 yr, the average loss rate has been much higher (~0.13 per cent/yr). (5)
Changes in transfer rate (5)	~100	
Changes in total energy (5)	10^3–10^4	
Secular Variation of Dipole Field		
Dipole "wobble" (eastward drift) (6)	1200–1800	Various estimates
Field strength oscillations (7)	~9000	Fundamental frequency
Polarity Reversals		
Change in field direction (8, 9, 10)	1000–4000	(8): intensity time scale ~10 times the direction time scale
Change in intensity (8, 9)	3000–10,000	(9): both time scales the same (3500 yr in two cases)
Interval between reversals (11)	0.03–30×10^6	2–3×10^5 yr over the last 50 m yr
	75×10^6	Solar vibration $\perp$ galactic plane ?
Long-term periodicity (?) (12)	250×10^6	Galactic rotation ?
	700×10^6	?

FOOTNOTES TO TABLE 4.1 (full references given at the end of the chapter))

(1): Braginskii (1964b, 1970a, b, 1971, 1972); Currie (1968, 1973a); Acheson and Hide (1973).
(2): Braginskii (1965b, 1967, 1970b, 1971, 1972); Hide (1966); Rikitake (1966b); Malkus (1967, 1971); Stewartson (1967, 1971); Gans (1971); Hide and Stewartson (1972); Soward (1972); P. H. Roberts and Soward (1972); Acheson and Hide (1973). (3): *For early references see* Jacobs (1963, pp. 70–76), Rikitake (1966a, p. 83 and p. 109). *See also* Pudovkin and Ye Valuyeva (1967, 1972); Yukutake (1968a, b, 1972); James (1968, 1970, 1971); Honkura and Rikitake (1972); P. H. Roberts and Soward (1972); Moffatt (1973). (4): McDonald and Gunst (1968); Verosub and Cox (1971); Cox (1972); Jin (1973). (5): Verosub and Cox (1971). (6): Kawai and Hirooka (1967); Kovacheva (1969); Márton (1970); Cox (1972). *See also* Jacobs (1971); Pudovkin and Ye Valuyeva (1972). (7): Smith (1967); Braginskii (1970b, 1971, 1972); Cox (1972). (8): Dunn *et al.* (1971). (9): Opdyke *et al.* (1973). (10): Harrison and Somayajulu (1966); Bullard (1968); Creer and Ispir (1970); Cox (1972). (11): Bullard (1968); Heirtzler *et al.* (1968); Helsley and Steiner (1969); McElhinny (1971); Vogt *et al.* (1972); Helsley (1972); Blakely and Cox (1972); Stewart and Irving (1973); Reid (1972). (12): Crain *et al.* (1969); Crain and Crain (1970); Ulrych (1972).

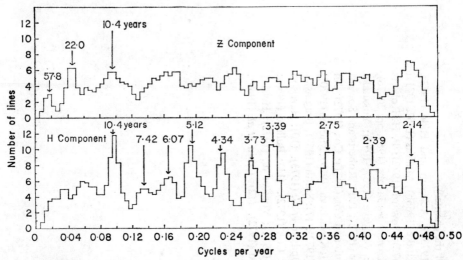

Fig. 4.6. H and Z smoothed histograms showing the distribution of a number of inter-polated lines as a function of frequency which occurred in 97 spectra. The "lines" were chosen by the computer without regard to signal to noise ratio. (After Currie, 1973a.)

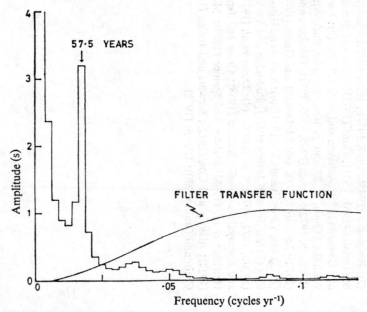

Fig. 4.7. Amplitude spectrum in seconds of time for yearly fluctuations in the length of the day from 1824–1950. Two hundred estimates were computed with a bandwidth of 0·0025 cycles yr^{-1} but only estimates up to 0·25 the Nyquist limit are plotted. Also shown is the transfer function of the high pass filter applied to the source data. (After Currie, 1973b.)

of a $50 \sim 60$ yr magnetic variation had been made earlier in connection with variations in the length of the day (l.o.d.)—e.g. Vestine and Kahle (1968) (see Section 4.8). The curves of Orlov (1965) based on data from four observatories, clearly show a $50 \sim 60$ yr cycle. However, until the more detailed and conclusive analysis of Currie (1973a), such a variation does not appear to have been widely known or accepted. In a further paper (1973b), Currie applied the maximum entropy method to 19th and 20th century data on variations in the l.o.d. The spectrum is dominated by a strong line at 57·5 yr (see Fig. 4.7) which is close to the estimate of 57·8 yr in the magnetic spectrum (Fig. 4.6). Currie suggested that the ~ 60 yr lines in the geomagnetic and astronomical data are causally related and that these results are the spectral analogues of the correlation previously established between l.o.d. variations and the westward drift of the eccentric dipole (Vestine and Kahle, 1968; Kahle *et al.*, 1969).

In 1839 Gauss showed that the field of a uniformly magnetized sphere, which is the same as that of a dipole at its centre, is an excellent first approximation to the Earth's magnetic field. Gauss further analysed the irregular part of the Earth's field, i.e. the difference between the actual observed field and that due to a uniformly magnetized sphere, and showed that both the regular and irregular components of the Earth's field are of internal origin. The geomagnetic poles, i.e. the points where the axis of the geocentric dipole which best approximates the Earth's field meets the surface of the Earth, are situated at approximately 79°N, 70°W, and 79°S, 110°E. The geomagnetic axis is thus inclined at about 11° to the Earth's geographical axis. If greater accuracy is needed, the potential of the Earth's magnetic field may be expanded in a series of spherical harmonics. It can also be shown that a better approximation to the Earth's field can be obtained by displacing the centre of the equivalent dipole by about 300 km in the direction of Indonesia.

A brief description of the geomagnetic field at the surface and inside the Earth is given in Table 4.2. The non-dipole component of the Earth's field, though much weaker than the dipole component, shows more rapid changes. The time scale of the non-dipole changes is measured in decades and that of the dipole in centuries. The isoporic foci also drift westward at a fraction of a degree per year. The drift of the dipole field is slower than that of the non-dipole field, at least by a factor of three. Both eastward and westward drifts of the non-dipole field have been inferred from palaeomagnetic studies. Observatory records from Sitka, Alaska, indicate an eastward drift during the past 60 yr, in contrast to the predominance of a westward direction of drift observed over most areas of the world for the past several hundred yr (Skiles, 1970; Yukutake, 1962). In addition to the westward drift, the pattern of the secular variation field may alter appreciably in a few decades. In some areas the amplitude of the secular change is anomalously large or small. At present

Table 4.2. Geomagnetic fields of internal origin. (After Gilliland 1973.)

Location	Strength $(\Gamma = 10^{-4}\ T)$	Spatial structure	References
Surface			
Dipole field (at magnetic poles)	0·6	Mainly dipolar, with dipole axis inclined about 11° to the axis of rotation at the present time. Field averages to an axial dipole within about $2 \cdot 5 \times 10^4$ yr, and to a geocentric axial dipole within about 2×10^6 yr.	McElhinny and Evans (1968); Opdyke (1972); P. H. Roberts and Soward (1972)
Non-dipole field	0·02	Average field has been an axial dipole for at least 2–3×10^9 yr.	
Core–mantle interface			
Poloidal (extrapolation of surface fields)	5–6 (?)	Probably mainly poloidal, with the non-dipole field of the same order as, or greater than, the dipole field.	Lowes (1972) P. H. Roberts and Soward (1972)
Toroidal (estimate)	0·2	Small toroidal component due to non-zero conductivity of lower mantle.	The toroidal field was estimated by making use of expressions given by Rochester (1960)
Core (various estimates)	50–500	Probably mainly toroidal	Bullard and Gellman (1954); Hide (1966); Braginskii (1971); P. H. Roberts and Soward (1972); Busse (1973b).

it is larger than average in the Antarctic and smaller than average in the Pacific hemisphere.

4.2. The Origin of the Earth's Magnetic Field

The problem of the origin of the Earth's magnetic field (and secular variation) is one of the oldest in geophysics and one to which no completely satisfactory answer has as yet been found. A number of possible sources for the field have been suggested—such as permanent magnetization or theories involving the rotation of the Earth—but most of them have proved to be inadequate. The only possible means seems to be some form of electromagnetic induction, electric currents flowing in the Earth's fluid, electrically-conducting core. This still poses the problem of how such currents were initiated—perhaps they arose from chemical irregularities which separated charges and set up a battery action generating weak currents. Palaeomagnetic measurements have shown that the Earth's main field has existed throughout geologic time and that its strength has never differed significantly from its present value. In a bounded, stationary, electrically conducting body, any system of electric currents will decay. The field or the current may be analysed into normal modes each of which decays exponentially with its own time constant. The time constant is proportional to σl^2 where σ is the electrical conductivity and l a characteristic length representing the distance in which the field changes by an appreciable amount. For a sphere the size of the Earth the most slowly decaygin mode is reduced to $1/e$ of its initial strength in a time of the order of 100,000 yr. Since the age of the Earth is more than 4000 m yr, the geomagnetic field cannot be a relic of the past, and a mechanism must be found for generating and maintaining electric currents to sustain the field. The most likely source of the electromotive force needed to maintain these currents is the motion of core material across the geomagnetic lines of force. The study of this process, in which the currents generated reinforce the magnetic field which gives rise to the driving e.m.f. is known as the homogeneous dynamo problem.

The dynamo theory of the Earth's magnetic field was due originally to Sir Joseph Larmor who in 1919 suggested that the magnetic field of the sun might be maintained by a mechanism analogous to that of a self-exciting dynamo. The pioneering work in dynamo theory was later carried out by Elsasser (1946a, b, 1947) and Bullard (1949a, b). The Earth's core is a good conductor of electricity and a fluid in which motions can take place, i.e. it permits both mechanical motion and the flow of electric current, the interaction of which could generate a self-sustaining magnetic field. It has not been possible to demonstrate the existence in the laboratory of such a dynamo action. If a bowl of mercury is heated from below, thermal convection will be set up—but no electric currents or magnetism will be detected in the bowl. Such a model experiment fails because electrical and mechanical processes do not scale down in the same way. An electric current in a bowl of mercury 30 cm in diameter would have a decay time of about one hundredth of a second. The decay time, however, increases as the square of the diameter—an electric current in the Earth's core would persist for about 10,000 yr before it decayed. This time is more than sufficient for the current and its associated magnetic field to be altered and amplified by motions in the fluid, however slow.

Even if energy sources exist within the Earth's core sufficient to maintain the field, there remains the critical problem of sign, i.e. it must also be shown that the inductive reaction to an initial field is regenerative. An engineering dynamo is exceedingly non-homogeneous, containing rotors, stators, wires, bearings, etc., whereas the fluid core of the Earth is simply-connected, and virtually homogeneous and isotropic. It might be suspected therefore that any dynamo trying to operate within such a fluid would be, in some sense, short-circuited; it is thus not at all clear that such a homogeneous dynamo can work under any conditions at all. In fact in 1934 Cowling showed that a magnetic field that possesses an axis of symmetry could not be produced by dynamo action. For many years, this important result was generally misinterpreted as implying also that an axisymmetric motion could not generate a magnetic field, but this has since been shown to be untrue under certain circumstances. The limitations imposed by Cowling's theorem (and its incorrect extension) led at one time to fears that dynamo action might in fact not be possible under any conditions: perhaps some "super-Cowling theorem" existed that prohibited it. Fortunately this has not proved to be the case. Cowling's result has however been extended by Backus and Chandrasekhar (1956) and it now appears that homogeneous dynamos must possess a low degree of symmetry. A number of other "non-dynamo" theorems that prohibit particular types of motion in a sphere from acting as dynamos have also been obtained. Some of these results are summarized in Tables 4.3 and 4.4. Most kinds of symmetry appear to be excluded. Some of these theorems restrict the types of possible motions and others restrict the field.

It is clear that the dynamics of the core is governed by Lorentz and Coriolis forces. It is usually supposed that the fluid motion contains some differential rotation which winds up any dipole field present into a toroidal field. Such a mechanism is very efficient and well understood, and presents a simple way in which the fluid may amplify the field—the field may be increased indefinitely simply by increasing the fluid motion. This does not solve the dynamo problem—to do that the dipole field must in some way be produced from the toroidal field, thus completing a cycle by which energy may be put into the field. This second stage of the cycle is much more difficult to account for. It is often assumed that a principal action of the Lorentz force is to limit the

Table 4.3. Anti-dynamo theorems for stationary magnetic fields.

Pairs of fields (v, H) with the following characteristics cannot be solutions to the kinematic dynamo problem, if it is required that $\partial H/\partial t = 0$.

Velocity v	Magnetic field H	References
Arbitrary	Axisymmetric	Cowling (1934, 1957, 1968)
Arbitrary	Two-dimensional	Cowling (1957); Lortz (1968)
Arbitrary	Poloidal	see Childress (1968)
Arbitrary	Toroidal	see Childress (1968)
Radial, vanishing at boundary	Arbitrary	Namikawa and Matsushita (1970)

Table 4.4. Anti-dynamo theorems for general magnetic fields.

Pairs of fields (v, H) with the following characteristics cannot be solutions to the kinematic dynamo problem.

Velocity v	Magnetic field H	References
Toroidal (in a sphere)	Arbitrary	Bullard and Gellman (1954); Cowling (1957)
Axisymmetric	Axisymmetric	Backus and Chandrasekhar (1956); Backus (1957); Cowling (1957); Braginskii (1965a)
Pure sin mϕ or cos mϕ dependence about axis of H symmetry	Nearly axisymmetric	Braginskii (1965b); P. H. Roberts (1967b).
Slow (magnetic Reynolds number for mean flow <1)	Arbitrary	Childress (1968); P. H. Roberts (1967a, b).
Insufficient rates of strain ($\partial u_i / \partial x_j$ too small)	Arbitrary	Backus (1958); Childress (1968).
Arbitrary, in a bounded, simply connected, perfectly conducting medium surrounded by a nonconductor.	Extending into the non-conducting medium	Bondi and Gold (1950); Leorat (1969).

magnitude of differential rotation, rather than suppressing other motions, as supposed in some recent dynamical theories. It is now known that small scale velocities can lead to dynamo action (as periodic motions or turbulence) and the small scale magnetic fields produced by them may contribute to a large Lorentz force. Busse (personal communication, 1973b) has further suggested that a balance between Coriolis and Lorentz forces may not hold in the core, and that most of the Coriolis force is balanced by the pressure. Such arguments have weakened the case for having a large ($\simeq 100\Gamma$) toroidal field in the core.

Several successful dynamos have been developed that involve axially symmetric motions, the earliest of which are those of Tverskoy (1966) and Gailitis (1970). Their models are interesting in that the velocity field is axisymmetric, although the resulting magnetic fields are not: the symmetry restrictions of Cowling's theorem are thus avoided.

In 1958 two dynamo models were found. However, though rigorous mathematical solutions were obtained, in each case the motions were physically very improbable. The first model was due to Backus (1958) who made use of the fact that the natural decay time of field components in a stationary conductor depends upon the length-scale of the component. He proposed a model in which short periods of vigorous motion within a fluid sphere are followed by longer periods with no motions. During these longer periods, all field harmonics except that of the lowest degree, and therefore greatest characteristic length, diffuse away. Another period of non-zero motion strengthens this harmonic and twists it into a different direction, and after several repetitions of this process the original field is regenerated.

The second model, due to Herzenberg (1958) consists of two spheres in the core each of which rotates as a rigid body at a constant angular velocity about a fixed axis. The axially symmetric component of the magnetic field of one of the spheres is twisted by rotation resulting in a toroidal field which is strong enough to give rise to a magnetic field in the other sphere. The axial component of this field is twisted as well and fed to the first sphere. If the rotation of the spheres is sufficiently rapid, a steady state may be reached. Lowes and Wilkinson (1963) built a working model of what is effectively a homogeneous self-maintaining dynamo based on Herzenberg's theory. For mechanical convenience they used, instead of spheres, two cylinders placed side by side with their axes at right angles to one another so that the induced field of each is directed along the axis of the other. If the directions of rotation are chosen correctly, any applied field along one axis will lead, after two stages of induction, to a parallel induced field. If the velocities are large enough the induced field will be larger than the applied field which is no longer needed i.e. the system would be self-sustaining. Rikitake and Hagiwara (1966) have investigated the stability of a Herzenberg dynamo and concluded that it i unstable for small disturbances about its steady state. Their analysis is not applicable to the case of the Earth however since numerical integrations could only be performed for parameters very different from those in the Earth's core.

A different type of model has been suggested by Parker (1955). His velocity field has two components—rotation (producing a toroidal field from a poloidal field) and convection (producing a poloidal field from a toroidal field). He supposed that a rapidly rotating fluid sphere is heated from within, leading to large-scale, asymmetric convection cells. As fluid in these cells rises and falls, the conservation of angular momentum results in non-uniform

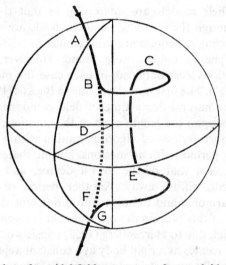

Fig. 4.8. The production of toroidal field components from poloidal field components by differential rotation: the poloidal field line ADG is stretched by differential rotation to produce the toroidal components BC and EF. (After Elsasser, 1950.)

rotation of the fluid, with the areas further away from the axis rotating more slowly than those nearer. This rotational shear will draw out magnetic lines of force that are in the meridionial plane (ADG in Fig. 4.8), to produce an azimuthal component of the field (BC and EF in Fig. 4.8). Eventually the magnetic stresses that develop will become large enough to reduce the differential rotation, and the stretching of the field lines will cease. The powerful toroidal field that results from such a mechanism can never be observed at the surface of the sphere, as no portion of it can escape from the fluid (i.e. it is a "contained" field). Parker next suggested that the action of the Coriolis force on the convection cells produces a cyclonic or twisting motion within these cells, rather like the circular motions in atmospheric weather systems. The convective lifting and rotational twisting in the cells between them turn the toroidal lines of force into a loop of flux (see Fig. 4.9, right-hand portion). This loop includes the new poloidal field components AB and CD in its projection on to the meridional plane (Fig. 4.9, left-hand portion). With diffusion outward of the component CD and inward of the component, AB, and continuous coalescence of such flux lines produced in many of these cyclonic convection cells, Parker suggested that it is possible that the original poloidal field could be regenerated and maintained.

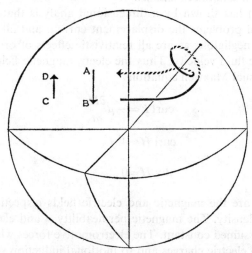

Fig. 4.9. The production of poloidal field components from toroidal field components by cyclonic twisting: the convective lifting and rotational twisting motion turns the toroidal field line into a loop (right hand half of the figure) which includes the new poloidal field components AB and CD in its projection on the meridional plane (left hand half of the figure). (After Frazer, 1973b.)

4.3. The Homogeneous Dynamo Equations

The homogeneous dynamo problem involves the solution of a highly complicated system of coupled partial differential equations. A good account of the derivation of these equations has been given by P. H. Roberts (1967a). The equations fall into four major groups.

(i) *The electrodynamic equations.* These include Maxwell's equations, the constitutive relations among the various electric and magnetic fields, Ohm's law, and the transformation relating the fields observed in one reference frame to those observed in another in relative motion.

(ii) *The hydrodynamic equations.* These include the equations of the conservation of mass and the conservation of momentum, and the constitutive equation for the total stress tensor.

(iii) *The thermodynamic equations.* These include the postulate of local thermodynamic equilibrium, the equation of heat conduction, and the constitutive law for the heat conduction vector. When combined with (i) and (ii) above, these equations lead to a detailed description of the energy flow within the system considered.

(iv) *The boundary and initial conditions.*

Elsasser (1954) has shown by a dimensional analysis that in geophysical and astrophysical problems the displacement current and all purely electrostatic effects are negligible, as are all relativistic effects of order higher than v/c where v is the fluid velocity. Thus the electromagnetic field equations (in e.m.u.) are the usual Maxwell equations

$$\text{curl } E = -\mu \frac{\partial H}{\partial t} \tag{4.1}$$

$$\text{curl } H = 4\pi j \tag{4.2}$$

$$\text{div } H = 0 \tag{4.3}$$

where H and E are the magnetic and electric fields respectively, and j the electric current density. The magnetic permeability μ and electrical conductivity σ will be assumed constant. The electromotive forces which give rise to j are due both to electric charges and to motional induction so that the total current j is given by

$$j = \sigma (E + \mu v \times H). \tag{4.4}$$

Taking the curl of Eqn. (4.2) and using Eqns. (4.4) and (4.1), E can be eliminated, leading to the equation

$$\text{curl curl } H = 4\pi\mu\sigma (-\partial H/\partial t + \text{curl } v \times H). \tag{4.5}$$

Since curl curl $H = \text{grad div } H - \nabla^2 H = -\nabla^2 H$ on using Eqn. (4.3) we finally obtain

$$(\partial H/\partial t) = \text{curl } (v \times H) + v_m \nabla^2 H \tag{4.6}$$

where

$$v_m = 1/4\pi\sigma\mu \tag{4.7}$$

is the "magnetic diffusivity".

Equations (4.3) and (4.6) give the relations between H and v which have to be satisfied from electromagnetic considerations. The term $\nabla^2 H$ causes the field to decay and, for a dynamo, it is essential that this term should be balanced and the decay prevented by the term curl $(v \times H)$ which represents the interaction between the velocity and the field.

If the material is at rest, Eqn. (4.6) reduces to

$$(\partial H/\partial t) = v_m \nabla^2 H. \tag{4.8}$$

This has the form of a diffusion equation, and indicates that the field leaks through the material from point to point. Dimensional arguments indicate a decay time of the order L^2/v_m where L is a length representative of the dimensions of the region in which current flows. For conductors in the laboratory this decay time is very small—even for a copper sphere of radius 1 m it is less than 10 sec. For cosmic conductors, on the other hand, because of their enormous size, it can be very large. Elsasser (1954) has estimated the time of free decay of the Earth's field to be of the order of 15,000 yr, supposing the core to consist of molten iron.

As an alternative limiting case, suppose that the material is in motion but has negligible electrical resistance. Equation (4.6) then becomes

$$(\partial H/\partial t) = \text{curl } (v \times H). \tag{4.9}$$

This equation is identical to that satisfied by the vorticity in the hydrodynamic theory of the flow of non-viscous fluid where it is shown that vortex lines move with the fluid. Thus Eqn. (4.9) implies that the field changes are the same as if the magnetic lines of force were "frozen" into the material.

When neither term on the right-hand side of Eqn. (4.6) is neglibible, both the above effects are observed, i.e. the lines of force tend to be carried about with the moving fluid and at the same time leak through it.

If L, T, V represent the order of magnitude of a length, time and velocity respectively, transport dominates leak if $LV \gg v_m$. The condition for the onset of turbulence in a fluid is that the non-dimensional Reynolds number $R_e = LV/v$ be numerically large. By analogy, a magnetic Reynolds number R_m may be defined as

$$R_m = LV/v_m. \tag{4.10}$$

Thus the condition for transport to dominate leak is that $R_m \gg 1$. This condition is only rarely satisfed in the laboratory—in cosmic masses, however, it is easily satisfied because of the enormous size of L. Thus under laboratory conditions, lines of force slip readily through the material—in cosmic masses, on the other hand, the leak is very slow and the lines of force can be regarded as very nearly frozen into the material.

To the electromagnetic equations, must be added the hydrodynamical equation of fluid motion in the Earth's core (the Navier–Stokes equation) together with the equation of continuity, which, for an incompressible fluid (the speed of flow v is much less than the speed of sound in the Earth's core) reduces to

$$\text{div } v = 0. \tag{4.11}$$

The Navier–Stokes equation is

$$\rho \left\{ \frac{\partial v}{\partial t} + (v \cdot \nabla) \, v + 2\Omega \times v - \nu \nabla^2 v \right\} - \frac{\mu}{4\pi} \text{ curl } H \times H = -\nabla p + \rho \nabla W \tag{4.12}$$

where v is the velocity relative to a system rotating with angular velocity Ω, p the pressure, W the gravitational potential (in which is absorbed the centrifugal force) and ρ and ν, the density and kinematic viscosity respectively. Equations (4.6) and (4.12) contain only the vectors v and H and are the basic equations of field motion.

Because of the complexity of the equations describing the hydromagnetic conditions in the Earth's core, most effort has been directed to seeking solutions of Maxwell's equations for a given velocity distribution. This approach, known as the kinematic dynamo problem is linear and has been the subject of much investigation. Expansion in spherical harmonics reduces Eqn. (4.6) to an infinite set of differential equations containing the radial functions for each harmonic, their first and second radial derivatives and their first time derivatives. For a given set of initial conditions these could in theory be integrated in time steps. Little progress has been made in this direction however and most workers have looked for steady state solutions, putting $\partial H/\partial t = 0$. In 1954, Bullard and Gellman looked for steady state solutions using first order finite differences for the radial variation. When the series are truncated a generalized algebraic eigenvalue problem in R_m remains. The difficulty has been in obtaining a convergent solution when the truncation level is raised. Bullard and Gellman (1954) chose as their velocity function differential rotation T_1 and a poloidal convection P_2^{2c} and obtained steady state solutions which appeared satisfactory. However with the advent of larger computers Gibson and P. H. Roberts (1969) were able to show that in fact their solution did not converge. A clue to the reason for the failure of the Bullard–Gellman dynamo can be found in the work of Braginskii (1965a, b).

Braginskii considered the generation of magnetic fields by a fixed fluid velocity that is mainly toroidal and axisymmetric, but which also has a small asymmetric component. Under certain limiting conditions (primarily a high value of the magnetic Reynolds number) he showed that a necessary condition for dynamo action was the simultaneous existence of both $\sin m\phi$ and $\cos m\phi$ terms, with the same value of m in each, in the Fourier expansion of the asymmetric velocity component (ϕ being the azimuthal angle in spherical polar co-ordinates r, θ, ϕ). Unless both of these terms occur, dynamo action is not possible. The velocity distribution used by Bullard and Gellman contained only a $\cos 2\phi$ term, with no $\sin 2\phi$ term, and this may be the reason for its failure. Lilley (1970) added a term in $\sin 2\phi$ to their velocity distribution, i.e. he assumed a velocity field of the form

$$v = T_1 + P_2^{2c} + P_2^{2s}$$

and demonstrated convergence which appeared to be much better than Bullard and Gellman's. Later work (P. H. Roberts, 1972; Gubbins 1973) however has shown that the convergence is still not satisfactory. Gubbins (1973) has since attempted to find solutions using modifications of Lilley's velocity field, but without success. A possible reason for the failure of Lilley's model may be that steady state solutions do not exist. His velocity field contains a non-zero angular rotation and the field may rotate—such a rotating field would not yield a solution to the steady state eigenvalue problem.

The first satisfactory numerical demonstration of dynamo action in a sphere was given by G. O. Roberts (see P. H. Roberts, 1971b). A solution was possible because he chose an axially symmetric velocity, for which solutions of the induction equation proportional to $e^{im\phi}$ decouple. G. O. Roberts considered the case $m = 1$, the calculation being essentially two dimensional. In his analysis he used a finite difference method rather than spherical harmonics. Flows similar to those of Roberts have also been investigated by Gubbins (1972, 1973) and Frazer (1973a) using the Bullard–Gellman expansion method. Gubbins (1973) in particular obtained dynamo action in which the numerical convergence is convincing. P. H. Roberts and Kumar have reported in Moffatt (1973) apparent convergence for the velocity field

$$v = T_1^0 + P_2^0 + P_2^{2s} + T_3^{2s}.$$

If convergence can be satisfactorily demonstrated, this will be the first realistic three dimensional numerical dynamo model.

G. O. Roberts (1970, 1972) has also investigated dynamos in infinite bodies of fluids. In particular he has shown that "almost all" spatially periodic motions in an unbounded conductor will lead to dynamo action. These dynamos are remarkable in that, although the motion extends to infinity, the field does not. Also the field, unlike the motion, is not spatially periodic. The real significance of these interesting and unexpected results is not yet clear. Childress (1967, 1968, 1970) has been able to show that such a periodic motion giving dynamo action in an unbounded conductor retains this property when fitted into a finite spherical volume by means of a "cut off" function.

4.4. Mean Field Electrodynamics

Most of the above dynamo models use large-scale, highly ordered fluid motions—i.e. motions in which the characteristic length of the velocity field is not much less than the radius of the sphere. In the early 1950's several attempts were made to produce models in which turbulent (i.e. random and small-scale) velocities might act as dynamos. Recent developments seem to have overcome many of the difficulties of the initial work in this direction. The modern theory, which has been called mean field electrodynamics, has been developed independently by Moffatt (1970) in Britain and by F. Krause, K.-H. Rädler and M. Steenbeck in Germany. A translation of the relevant work of the German school on turbulent dynamos has been made by P. H. Roberts and Stix (1971). The scope of these investigations goes far beyond the terrestrial dynamo and is important in the general field of cosmical electrodynamics.

In mean field dynamo models the velocity v and magnetic field H are each represented as the sum of a statistical average and a fluctuating part. The average fields $\bar{v}$ and $\bar{H}$ are assumed to vary on a length scale L, while the fluctuating fields v' and H' (with zero statistical average) are assumed to vary on a length scale l ($l \ll L$).

The statistical average of Ohm's law for a moving medium (i.e. the average of Eqn. (4.4)) is

$$\bar{j} = \sigma(\bar{E} + \mu\bar{v} \times \bar{H} + \mu\overline{v' \times H'}). \tag{4.13}$$

This equation contains a "new" electromotive force $\mu\overline{v' \times H'}$. If this e.m.f. can be represented as a functional of the mean fields $\bar{v}$ and $\bar{H}$, the mean field kinematic dynamo problem becomes closed. Considerable attention has been focussed on the derivation of simple representations for $\overline{v' \times H'}$. Parker (1955) drew attention to the possibility that

$$\overline{v' \times H'} = \alpha\bar{H}.$$

Steenbeck and Krause (1966) have christened this the α-effect. Several successful α-effect dynamos have been studied both for the case $\bar{v} = 0$ and for the case $\bar{v} \neq 0$.

If the turbulence is isotropic (as well as homogeneous) then its statistical properties are not changed by any symmetry transformations. In other words, moving or rotating the axes, or viewing the system through a mirror, does not change the appearance or the mathematical description of the turbulent velocity. In this case, the critical tensor* is zero, and so, on the average, there is no electromagnetic induction, i.e. isotropic turbulence cannot maintain a large-scale magnetic field through dynamo action.

However, if we remove the condition of reflectional symmetry from the turbulence, the components of the critical tensor take on non-zero values, so

---

* v' and H' can be expressed in terms of Fourier integrals and a critical tensor (or matrix) determined by the spectrum of the kinetic energy of the turbulence.

that in this case the turbulence does affect the large-scale field. The field can in fact be maintained by this turbulence, with the rate of generation of the field depending upon a quantity, which Moffatt terms the "helicity" defined as $\overline{v' \cdot \text{curl } v'}$ the bar again meaning an average. We can picture the meaning of helicity by representing a turbulent velocity flow by a box containing a large number of randomly directed screws. If the spiral edge of each screw represents the short-term path of a moving element of fluid, then the fluid over-all will have zero helicity if there are equal numbers of right-handed (i.e. normal) thread screws and left-handed (i.e. reversed) thread screws. In such a case, viewing the fluid—or the box of screws—through a mirror produces no visible change in the average properties: turbulence with zero helicity has reflectional symmetry. On the other hand if more screws in the box are right-handed than are left-handed (or vice versa), the helicity is non-zero, and viewing the system in a mirror clearly produces a different average picture, interchanging a surplus of right-handed screws for a surplus of left-handed ones (or vice versa). Thus helicity measures the lack of reflectional symmetry in the turbulence, and it is turbulence that lacks such symmetry that can give rise to dynamo action. This approach to the dynamo problem has been extended by Moffatt (1972) to include the equations of motion, leading towards a full dynamical theory.

4.5. Reversals of the Earth's Magnetic Field

One of the most interesting results of palaeomagnetic studies is that many igneous rocks show a permanent magnetization approximately opposite in direction to the present field. Reverse magnetization was first discovered in 1906 by Brunhes in a lava from the Massif Central mountain range in France —since then examples have been found in almost every part of the world. About one-half of all rocks measured are found to be normally magnetized and one-half reversely. Dagley et al. (1967) carried out an extensive palaeomagnetic survey of Eastern Iceland sampling some 900 separate lava flows lying on top of each other. The direction of magnetization of more than 2000 samples representative of individual lava flows was determined covering a time interval of 20 m yr. At least 61 polarity zones, or 60 complete changes of polarity were found giving an average rate of at least 3 inversions/m yr.

There is no à priori reason why the Earth's magnetic field should have a particular polarity, and there is no fundamental reason why its polarity should not change. It is easy to see that dynamos can produce a field in either direction. Equation (4.6) is linear and homogeneous in the field and Eqn. (4.12) inhomogeneous and quadratic. Thus if a given velocity field will support either a steady or a varying magnetic field, then it will also support the reversed field and the same forces will drive it. This, however, merely shows that the reversed field satisfies the equations—it does not prove that reversal will take place. It is possible that some physical or chemical processes exist whereby a material could acquire a magnetization opposite in direction to that of the ambient field. In fact Néel (1951, 1955) suggested, on theoretical grounds, four possible mechanisms—and within two years two of them had been verified, one by Gorter for a synthetic substance in the laboratory (although no naturally occurring rock has been found to behave in the required manner), and one by T. Nagata for a dacite pumice from Haruna in Japan. However the great majority of igneous rocks which show reversal in the field, unlike the Haruna pumice, do not exhibit this property in the laboratory.

To prove that a reversed rock sample has become magnetized by a reversal of the Earth's field, it is necessary to show that it cannot have been reversed by any physico-chemical process. This is almost impossible to do since physical changes may have occurred since the initial magnetization or may occur during laboratory tests. More positive results can only come from the correlation of data from rocks of varying types at different sites and by statistical analyses of the relation between the polarity and other chemical and physical properties of the rock sample.

The same pattern of reversals observed in igneous rocks has also been found in deep sea sediments (see e.g. Opdyke *et al.*, 1966). No two substances could be more different or have more different histories than the lavas of California and the pelagic sediments of the Pacific. The lavas were poured out, hot and molten, by volcanoes and magnetized by cooling in the Earth's field; the ocean sediments on the other hand accumulated grain by grain by slow sedimentation and by chemical deposition in the cold depths of the ocean. If these two materials show the same pattern of reversals then it must be the result of an external influence working on both and not due to a recurrent synchronous change in the two materials. The evidence seems compelling that reversals of the Earth's field are the cause of the reversals of magnetization, and this provides a further constraint on any theory of the origin of the field.

Reversals occurred during the Precambrian and have been observed in all subsequent periods. There is no evidence that periods of either polarity are systematically of longer or shorter duration. However during the Kiaman—a period of about 60 m yr within the upper Carboniferous and Permian (about 235–290 m yr ago) the polarity of the Earth's field appears to have been almost always reversed—until quite recently no normal intervals at all were known within this period.

The Earth's field reverses in a highly irregular fashion—so irregular that geomagnetic reversals are often modelled as a non-stationary random process. The frequency of geomagnetic reversals is plotted as a function of time in Fig. 4.10. The smooth, dashed curve has been taken from the work of McElhinny (1971), who plotted the percentage of polarity measurements that are "mixed" (i.e. both "normal" and "reversed" in the same rock unit) as a function of time. The curve is clearly a very crude approximation, but serves to illustrate the non-stationary character of the reversal process. The detailed results of Heirtzler, *et al.* (1968), and of Helsley and Steiner (1969), plotted on the right-hand side in Fig. 4.10, indicate that changes in the reversal frequency

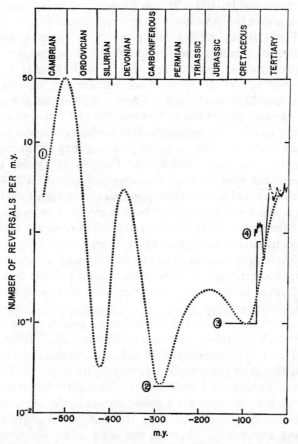

Fig. 4.10. Reversal rate of the geomagnetic field as a function of time. (1) Curve derived from the work of McElhinny (1971). McElhinny's plot of the percentage of "mixed" polarity measurements vs. time is interpreted, to a crude approximation, as a *logarithmic* plot of reversal rate vs. time. (2) Kiaman Magnetic Interval—approximately one reversal in 50 m yr. (3) Results of Helsley and Steiner (1969). (4) Results of Heirtzler *et al.* (1968), obtained from sea-floor anomaly patterns. (Plot shows a 10 m yr average taken every 1 m yr.)

are likely to be discontinuous on the time scale shown. There appear to be sudden jumps in the frequency of reversals at about 50 and 72 m yr ago.

Other results (not plotted in Fig. 4.10) indicate that the nonstationary character of the reversal process has persisted over much longer times than those indicated. Reid (1972) and Stewart and Irving (1973) found that reversal rates in the Precambrian varied in much the same way as those plotted for the Phanerozoic (the time range shown in Fig. 4.10). Reid reported a variation of the reversal rate from $0 \cdot 4/m$ yr to $1 \cdot 1/m$ yr over a 60 m yr interval roughly 1800 m yr ago. Stewart and Irving reported reversal frequencies less than $0 \cdot 1/m$ yr 990 m yr ago, and greater than $1/m$ yr 790 m yr ago. Since Fig. 4.10 was plotted, two papers have appeared which indicate that the peak shown in the Jurassic and Triassic is somewhat too low. Vogt *et al.* (1972) reported 41 reversals between 150 and 135 m yr ago, giving an average reversal rate of $2 \cdot 7/m$ yr in the late Jurassic, and Helsley (1972) found that at least 23 reversals occurred during the Triassic, giving an average reversal rate greater than $0 \cdot 7/m$ yr between 225 and 190 m yr ago. Detailed study of more recent palaeomagnetic data indicates that the time between reversals varies widely—from $\sim 30,000$ yr to as long as 30 m yr (Bullard, 1968; Heirtzler *et al.*, 1968; Blakely and Cox, 1972). During the last few million yr, reversals have occurred at intervals of roughly 200–300 thousand yr. Four major normal and reversed sequences have been found during the past $3 \cdot 6$ m yr. These major groupings have been called geomagnetic polarity epochs and have been named by Cox *et al.* (1964) after people who have made significant contributions to geomagnetism. Superimposed on these polarity epochs are brief fluctuations in magnetic polarity with a duration that is an order of magnitude shorter. These have been called polarity events and have been named after the localities where they were first recognized (see Fig. 4.11).

Several workers have succeeded in obtaining a record of the geomagnetic field during a polarity transition (see e.g. Watkins, 1967, 1969). It appears that during a reversal the intensity of the field first decreases by a factor of 3 or 4 for several thousand years while maintaining its direction. The magnetic vector then usually executes several swings of about 30°, before moving along an irregular path to the opposite polarity direction, the intensity still being reduced, rising to its normal value later. It is not certain whether the field is dipolar during a transition. There do not seem to be any precursors of a reversal or any indication later that a reversal has occurred. A detailed record of a field reversal has been described by Dunn *et al.* (1971). They found that the field intensity decreased by a factor of 10 before any change in field direction occurred, and did not return to normal until after the directional change was completed. The directional change was estimated to have taken 1000–4000 yr, while the intensity change took 10,000 yr.

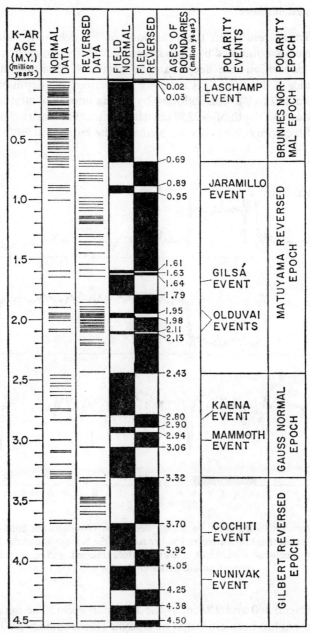

Fig. 4.11. Time scale for geomagnetic reversals. Each short horizontal line shows the age as determined by potassium–argon dating and the magnetic polarity (normal or reversed) of one volcanic cooling unit. Normal polarity intervals are shown by the solid portions of the "field normal" column, and reversed polarity intervals by the solid portions of the "field reversal" column. The duration of events is based in part on palaeomagnetic data from sediments and magnetic profiles. (After Cox, 1969.)

Contradictory results on the behaviour of the magnetic field during a
reversal have been obtained by Opdyke *et al.* (1973) from measurements made
on a high deposition rate deep-sea core. Figure 4.12 shows the magnetic
characteristics of the core—it is reversely magnetized to a depth of ~460 cm,
normal from 460 to 940 cm, followed by a long unbroken stretch of reverse
magnetization to a depth of ~2250 cm where a further polarity change occurs.
The rest of the core is normally magnetized. The authors identify the normal

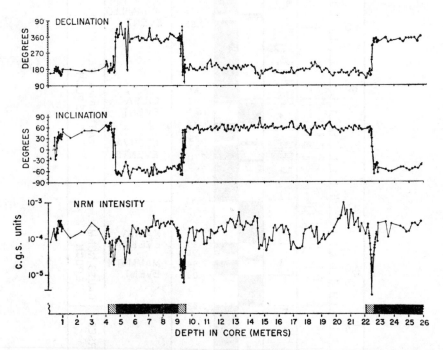

Fig. 4.12. Variations with depth in a high deposition rate deep sea core of the NRM
declination, inclination and intensity. The bar scale at the bottom shows the interpretation
of the NRM directional data: black is normal, white reversed, and hachured intermediate
magnetization. (After Opdyke *et al.*, 1973.)

interval between 460 and 930 cm as the Jaramillo event. The duration of the
Jaramillo event has been estimated by Opdyke (1969) to be 56,000 yr, making
the rate of sedimentation during the event 8·6 cm/1000 yr in this core.
Because of this high rate of deposition, intermediate directions of magnetization were observed within each transition as well as a sharp decrease in
intensity. Details of the behaviour of the magnetic field at the lower Jaramillo

reversal are shown on an expanded scale in Fig. 4.13. It can be seen that the pronounced drop in the intensity of magnetization is coincident with the onset of the directional changes, contrary to what has previously been thought. The time taken for both the intensity and direction of the field to reverse is ~4600 yr—again in contrast to the results of Dunn *et al.* (1971) who found

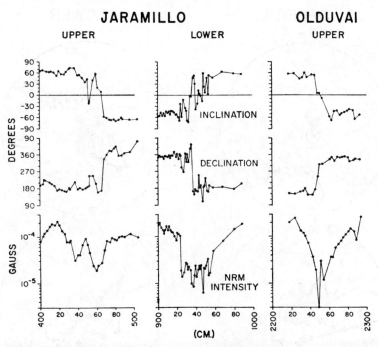

Fig. 4.13. Details of the inclination, declination and intensity of magnetically cleaned remanence across the upper and lower Jaramillo, and the upper Oldurai polarity transitions. (After Opdyke *et al.*, 1973.)

that, although the directional change occurred in ~4000 yr, the accompanying intensity variation took place over ~10,000 yr. Figure 4.13 also shows that, during the time of the polarity change, there are three cycles of intensity changes with periods ~1400 yr. The declination and inclination variations show a similar periodicity. The period of these movements is close to that usually associated with the secular variation of the non-dipole field, supporting the hypothesis that during a transition the dipole field is weak, being of comparable intensity to the non-dipole field.

Opdyke *et al.* also traced the movement of the virtual geomagnetic pole (VGP) during a reversal. Figure 4.14 shows the movements for the upper and lower Jaramillo transitions. For the lower transition the VGP describes first a clockwise and then a counter-clockwise loop in the southern polar regions. It then enters the northern hemisphere, tracing out a clockwise loop before settling down. These three loops are accompanied by the intensity variation described above.

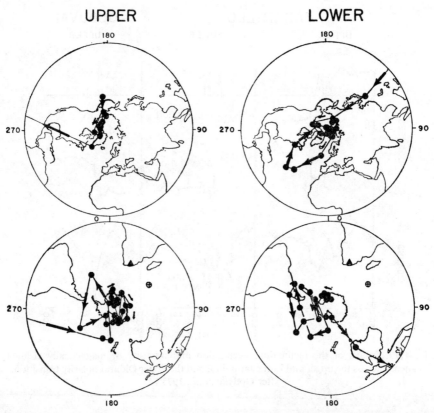

Fig. 4.14. Positions of the VGP for upper and lower Jaramillo polarity transitions. VGP's are calculated from cleaned remanent directions by suitable adjustment of declinations. Core location shown by circled cross. Position of VGP for intermediate direction from Jaramillo Creek, N. Mexico, dated at 0.86 m yr (Doell and Dalrymple, 1966) shown by solid triangle. (After Opdyke *et al.*, 1973.)

Creer and Ispir (1970) suggested that the field maintains dipolar characteristics during a polarity change and that the path of the dipole from one hemisphere to another passes through the Indian Ocean. None of the transition paths of Opdyke *et al.* pass through the Indian Ocean—the three reversed paths actually occur in different meridional quadrants. Their evidence does not therefore support the model of a toppling dipole: rather they believe that the intensity of the dipole field drops rapidly to a low, but non-zero, value, allowing the non-dipole field to predominate. This is reflected in the large looping excursions from the rotation axis seen in the VGP paths (Fig. 4.14). Since both clockwise and anticlockwise rotations of the VGP path are seen, it would appear that both westward and eastward drifts of the nondipole field occurred (Skiles, 1970). Both eastward and westward drifts of the non-dipole field have also been inferred from the sense of VGP loops associated with the upper Miocene Steens Mountain polarity change (Watkins, 1969) and those sedimentary cores from Lake Windermere covering the past 10,000 yr (Creer et al., 1972). Although the excursions of the VGP's are large, they only rarely extend into equatorial latitudes. This may indicate that although the dipole field is much reduced, it is still sufficient to keep the VGP's in middle and upper latitudes. During the lower Jaramillo reversal, the motion of the VGP from the southern to the northern hemisphere took place at the intensity minimum at the end of the second 1400 yr cycle. At this point, the dipole field must have changed its polarity very rapidly—in the order of a few hundred years.

Yaskawa (1973) has reported the results of a palaeomagnetic investigation of a 200 m core of deep sediments taken from Lake Biwa, Japan. The rate of sedimentation is estimated to be several hundred times greater than that of the sediments in the deep-sea bottom giving much more precise and detailed information on the time variation of the geomagnetic field. At least three short reversed events were found in the present Brunhes normal polarity epoch at intervals of about 90,000–100,000 yr. There may well have been in addition excursions of the geomagnetic pole. Barbetti and McElhinny (1972) have also reported excursions of the geomagnetic field from an analysis of ancient Australian fire places. Although the pattern of the pole–path movement is similar to that obtained from the core in Lake Biwa, the age of the Australian excursion is different from those found in Japan.

Kawai et al. (1973) have examined the time variation of the geomagnetic field during the Matuyama reversed polarity epoch, using a core from the North Pacific basin. The change of magnetic field with increasing depth from 25–197 cm is shown in Fig. 4.15. Above a depth of 52 cm, the declination was mostly normal, in the same direction as that of the present field. However, almost anti-parallel declinations were found to have occurred at least five times and seem to have lasted for only a short time. At a depth of 52 cm reversed declination suddenly appears and lasts until a depth of 74 cm. The boundary separating normal and reversed declinations is extremely sharp (it occurs within a 1 mm section of the core), suggesting that the change was far more rapid than the usually accepted time of ~4000 yr.

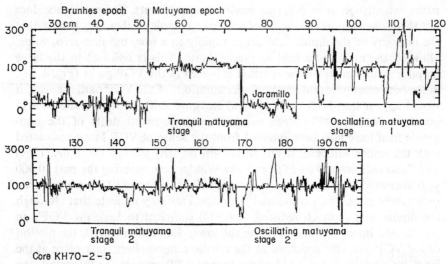

Fig. 4.15. Change of declination with depth in core from the North Pacific basin, showing the oscillation of the field during the Matuyama epoch. (After Kawai *et al.*, 1973.)

Between depths of 52–87 cm and 124–142 cm the declination is remarkably constant, except for the two sudden changes at the upper and lower Jaramillo transition zone. Kawai *et al.* call these periods "Tranquil Matuyama". During the rest of the Matuyama epoch there are strong oscillations of the magnetic field, although the total duration of reversed polarity is much longer than that of normal. Kawai *et al.* call this stage of the Matuyama epoch, "Oscillating Matuyama". They also investigated in detail the Brunhes–Matuyama boundary. They found a drop in intensity (to one quarter) prior to any directional change in the field. Subsequently the inclination began to decrease with as yet no change in the declination. The minimum in intensity lasted for about 500 yr, whilst the change in declination occurred ~2000 yr later. Kawai *et al.* speculated that perhaps a new field with opposite polarity existed from the time of the intensity drop and that perhaps two fields can co-exist for a long time. The "Tranquil Matuyama" period would then be the consequence of two approximately equal anti-parallel poloidal fields, whereas the "Oscillating Matuyama" regime is the result of an unbalance between the two fields—one fluctuating in intensity relative to the other.

Steinhauser and Vincenz (1973) have used data from 23 field reversals ranging in age from Recent to Upper Mesozoic to study the behaviour of the geomagnetic field during a polarity transition. In particular they investigated the longitudinal and latitudinal distribution of palaeopoles during a polarity change. They found two preferential meridional bands of polarity transition centred on planes through 40°E–140°W and 120°E–60°W. Both these bands are separated by two broad regions without transitional poles situated between 6°E–35°W and 85°E–70°W. The 40°E–140°W meridional band of preferential transitions is in agreement with the results of Creer (1972) who, using 12 polarity changes determined an average path centred on the plane through 60°E–120°W. As mentioned in Section 4.1, Doell and Cox (1971, 1972) found that the secular variation over the Central Pacific has been very smooth for at least the last 1 m yr and suggested that the non-dipole component of the main field has been very small in this region. It is significant that the Central Pacific region lies within the transition zone between 145°E and 174°W, i.e. within the band between 6°E and 35°W.

Steinhauser and Vincenz also investigated the latitudinal distribution of transitional palaeopoles. They found a U-shaped distribution, there being a decrease in the number of observed poles with decreasing latitude. They interpreted this result as reflecting an acceleration in the movement of the dipole axis when it approaches the equator—they estimated that it is moving 3·4 times faster in equatorial latitudes than in latitudes around 40°. In contrast the record of a non-dipole field would give a random latitudinal distribution of poles. They further estimated that the dipole moment is reduced by about one order of magnitude for only about 12 per cent of the transitional time, while for two-thirds of the time its magnitude is comparatively high—with field intensities considerably greater than the intensity of the non-dipole field.

Wilson *et al.* (1972) have carried out statistical analyses of the behaviour of the palaeomagnetic field during the past 12 m yr using the large amount of data available from Eastern and Western Iceland. They found that many lavas showed palaeomagnetic fields deviating greatly from the axial dipole field. These anomalously directed fields were also found to be significantly weaker than the more usual "normal" and "reversed" fields. Assuming that, even when in an anomalous state, the field is due to a non-axial centred dipole, it is possible to transform an ancient field strength into a pseudo dipole moment. Figure 4.16 shows, for the combined Eastern and Western Icelandic data, the variation of dipole moment with co-latitude θ. This figure provides a natural definition of normal and reversed states. Wilson *et al.* took the range $0 \leqslant \theta \leqslant 40°$ to define normal states, $40° < \theta \leqslant 140°$ to define intermediate states, and $140° < \theta \leqslant 180°$ to define reversed states. The normal state appears to have been more stable (56 per cent of the time) over the past 12 m yr, than the reversed state. The dependence of the time-averaged dipole strength on dipole orientation provides some constraint on any postulated mechanism for field generation, although it does not give any direct insight into the physical behaviour of the core. Wilson *et al.* speculated that perhaps some critical coupling within the core, necessary to maintain dynamo action producing normal and reversed states, begins to break down when the dipole tilts more than 40° from the rotation axis.

In a later paper, Dagley and Lawley (1974) re-examined 23 transition intervals in some detail, using the definition of Wilson *et al.* (1972) for normal and reversed states. They attempted to decide between two basic models which have been suggested to describe the behaviour of the geomagnetic field

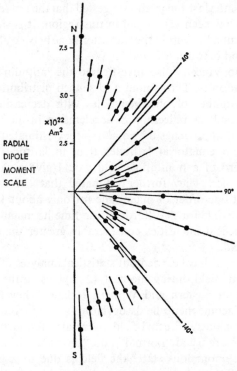

Fig. 4.16. Eastern and Western Icelandic data, separately normalized to the mean dipole moment ($5\cdot35 \times 10^{25}$ gauss cm³) and then combined in 5° intervals of VGP colatitude plotted in polar coordinates. (After Wilson *et al.*, 1972.)

during a polarity transition. In one model the main dipole field remains axial but decreases in strength during the transition so that the non-dipole field becomes dominant in the intermediate stages (see for example Larson *et al.*, 1971). In the other model the non-dipole field remains a small part of the total field but the dipole is considered to have two or three components the strength, polarity and possibly orientation of which change during the transition (Creer and Ispir, 1970). Creer and Ispir discuss their results in terms of a model field consisting of three dipoles. Bochev (1969) had already suggested such a model, arguing that such a representation gave more physical insight into the source of the geomagnetic field than the conventional expression in terms of a series of geocentric multipoles obtained from a spherical harmonic analysis. For epoch 1960, Bochev found that two dipoles are approximately parallel to the axis of rotation whilst the third is aligned obliquely to the Earth's axis. Creer and Ispir suggested that these three dipoles correspond to different dynamo processes and that each dynamo might drift or undergo periodic changes in position—each might also fluctuate in strength and even reverse its polarity independently of the others. Such a model, they claim, can explain a number of the more complex reversal paths.

Dagley and Lawley (1974) concluded that it is not possible to choose between the two models with the present amount of available data, although they do come to a number of conclusions. The diversity of pole paths and common westerly trends lend support to the model in which then on-dipoles field becomes dominant during the transition. However, certain similarities and the sharp east-west changes of longitude support a purely dipole model with an independently inverting equatorial dipole but there is no common pole path nor even a preferred sector of the globe as suggested by Creer and Ispir. Neither model is preferred and a combination of a three-component dipole and non-dipole fields may be needed. The very low magnetic inductions inferred from samples of intermediate polarity suggest that if the non-dipole field does persist when the axial dipole field becomes small then it too decreases in strength. It is not possible to say whether the non-dipole field predominates at any time during the transition. A possible link between inversions and tectonic activity is suggested by the detail recorded in some transitions while the fact that there are more transitions from the reversed mode of the field than from the normal mode implies that the two modes are not equally stable.

A number of models have been suggested to "explain" reversals of the Earth's field. Cox (1968) developed a probabalistic model in which it is assumed that polarity changes occur as the result of an interaction between steady oscillations and random processes. The steady oscillator is the dipole component of the field and the random variations are the components of the non-dipole field. The random variations serve as a triggering mechanism that produces a reversal whenever the ratio of the non-dipole to dipole fields exceeds a critical value (Fig. 4.17). Nagata (1969) suggested that the main dipole field is steadily maintained only as long as the convection pattern in the core is asymmetric, as is the case in the Bullard–Gellman–Lilley dynamo, but collapses when the convection pattern becomes symmetric as in the Bullard–Gellman model.

Parker (1969) developed a different model for reversals. He showed that a fluctuation in the distribution of the cyclonic convective cells in the core can

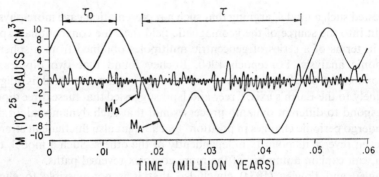

Fig. 4.17. Probabilistic model for reversals. t_d is the period of the dipole field and t the length of a polarity interval. A polarity change occurs whenever the quantity $(M_A + M_A')$ changes sign, where M_A is the axial moment of the dipole field and M_A' is a measure of the non-dipole field. (After Cox, 1968.)

produce an abrupt reversal of the geomagnetic field. The simplest fluctuation leading to a reversal is a general absence of cyclones below about latitude 25° for a time comparable to the lifetime (~ 1000 yr) of an individual cell. In both Cox's and Parker's models, cyclonic convection cells in the core produce reversals by a two-step mechanism. At any instant they are randomly distributed throughout the core: reversals occur when, through random processes, they arrive at certain critical configurations. However, there is a fundamental difference between the two models. In that of Parker (and also that of Nagata) the occurrence of a reversal depends only on the spatial distribution of the cyclones and not on the intensity of the dipole field. In the model of Cox the occurrence of reversals depends both upon the distribution of the cyclones and on the field strength of the cyclone disturbances (i.e. the non-dipole field) relative to the dipole field.

4.6. The Full Hydromagnetic Dynamo Problem

One disadvantage of the kinematic dynamo is that the final equilibrium value of the field cannot be found—the solutions continue to grow indefinitely. The hydromagnetic dynamo on the other hand is able to account for the equilibrium value of the field. Although it is now known that kinematic dynamos exist, solutions in which Maxwell's equations are solved for specified velocities are of limited geophysical interest since there is no guarantee that there exist forces in the Earth's core that can sustain them. Without a satisfactory theory to account for the driving force, the problem is not realistic and has only been "pushed one stage further back".

In a dynamical theory the velocities would be calculated from assumed forces—almost no work has been done on this more difficult non-linear problem. From a consideration of the order of magnitude of the terms in the Navier–Stokes equation (4.12) it can be shown that the inertial terms may be neglected and most probably the viscous forces. We must then have a balance between the Coriolis forces, the pressure gradient, the electromagnetic forces and the applied force. It can easily be verified that if there is no applied force, the electromagnetic force cannot be balanced by the other two terms, since neither can supply energy. The pressure term can be removed by taking the curl of the resulting equation—this leads to an expression from which the curl of the force can be found for any dynamo for which the velocity and the field are known, i.e. if we had a solution to Maxwell's equations for a specified velocity field, it is possible, in theory, to calculate the forces needed to drive the system. To solve the inverse problem with specified forces is, as already mentioned, very much more difficult.

The simplest model of a hydromagnetic dynamo is the disk dynamo driven by a constant couple, originally studied by Bullard (1955). It is known that coupled disk dynamos exhibit geomagnetic type reversals (Rikitake 1958; Allan 1958, 1962), and this is still the only theoretical background for reversal behaviour. Cook and P. H. Roberts (1970) have also studied the reversal properties of coupled disk dynamos, but it was not until 1972 that Malkus found that a single disk dynamo, under the right conditions, could also exhibit reversals. Disk dynamos are very far removed from fluid dynamos, but do give valuable information on the nature of the equations governing reversal behaviour.

A number of mechanisms have been suggested which might maintain fluid motions in the Earth's core and hence drive the dynamo—these will be discussed in the next section. Only two of them, thermal convection and the precession of the Earth, seem at all possible. A review of recent advances in dynamo theory has been given by Gubbins (1974).

4.7. Energetics of the Earth's Core

Motions in the Earth's electrically conducting fluid core which are the probable cause of the geomagnetic secular variation have time scales of the order of a few hundred years or less. Using seismic bounds on the kinematic molecular viscosity of the core and order of magnitude arguments about the eddy viscosity, Backus (1968) showed that at such short periods it is reasonable to consider core motions to consist of a boundary layer of Ekman–Hartmann type close to the MCB, and an interior free-stream motion where the viscosity and resistivity are zero.

Kahle *et al.* (1967) had attempted to determine the fluid velocity v near the surface of the core from a knowledge of the magnetic field and secular varia-

F

tion at the top of the core—which information may be obtained by extrapolation from observations made at the surface of the Earth. However Backus (1968) has shown that the goal which Kahle *et al.*, set themselves is in principle unattainable and that there is no unique solution to the problem. The non-uniqueness involves not only fine scale features, but also arbitrary functions whose scales may be of the order of the core's circumference. Kahle *et al.* obtained a unique solution only because they truncated the spherical harmonic expansion of *v*. Their truncated expansion does not fully reproduce the secular variation, i.e. reduce the r.m.s. difference between the measured and computed values to zero.

If magnetic diffusion is neglected (i.e. if the electrical conductivity is considered infinite) in the free stream in the core, the external geomagnetic field is completely determined by the fluid motion at the top of the free stream. Backus (1968) showed that it is thus possible to test whether there is *any* motion of a perfectly conducting core which will produce the observed secular variation. If the observed secular variation passes this test, Backus showed how to obtain explicitly all "eligible" velocity fields i.e. all velocity fields at the top of the free stream in the core which are capable of producing exactly the observed secular variation. Because of this multiplicity of eligible velocity fields, it is not possible to determine whether there is a westward drift of core fluid, or a latitude dependence of the westward drift of the core fluid at the top of the free stream (that the *magnetic field* itself shows a westward drift was established by Bullard *et al.* in 1950). Backus showed how it is possible to select from among all eligible velocity fields those which are of particular geophysical interest, such as that which, in a least squares sense, is most nearly a uniform rigid rotation (westward drift) about the geographic polar axis of the Earth or that which is most nearly a latitude dependent westward drift. Booker (1969) has used the methods outlined by Backus to obtain velocity components which are compatible with primarily latitude dependent westward drift; they are not compatible however with the velocity field obtained by Kahle *et al.* (1967).

One of the outstanding features of the velocity field of Kahle *et al.* is a marked downflow (convergence) under the central Pacific ocean and a marked upflow (divergence) under Africa. Using a completely different approach, Rikitake (1967) derived core motions on the assumption that the geomagnetic non-dipole field is created by an interaction between a strong toroidal magnetic field and convective motions. It is interesting that he obtained the same marked upflow and downflow under Africa and the central Pacific as did Kahle *et al.* (see Fig. 4.18), although the two studies were based on entirely different premises. What is even more interesting is that Lee and Uyeda (1965) found in the orthogonal representation (to third-order spherical harmonics) of 987 heat flow values, a high and a low situated almost exactly

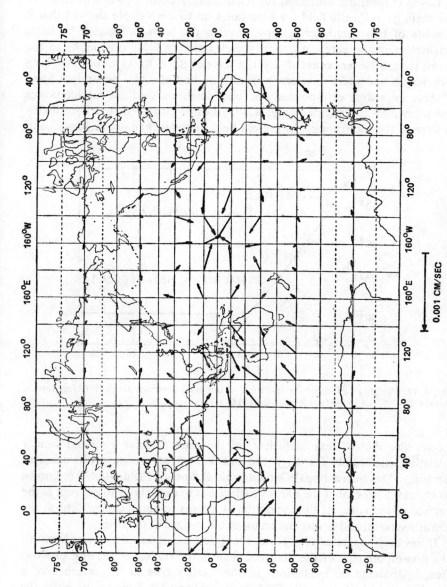

Fig. 4.18. Horizontal velocity at the surface of the core for epoch 1965. (After Rikitake, 1967.)

over the upflow under Africa and the downflow under the central Pacific. It is difficult to decide whether this correlation has any geophysical significance.

Lowes (1974a) has estimated the logarithmic spatial "power" spectrum of the main geomagnetic field for harmonics up to $n = 500$. He showed that it consists of two components, long wavelengths being dominated by fields originating in the core, and short wavelengths by fields originating in the crust; the cross-over occurs at $n \geqslant 11$, a wavelength $\leqslant 3600$ km. Both the long wavelength and short wavelength fields may be well fitted by straight lines. If these lines were extrapolated and combined, an idealized surface power spectrum may be obtained (Fig. 4.19)—Lowes estimated that it probably represents the spectrum of the present field to better than about 30 per cent.

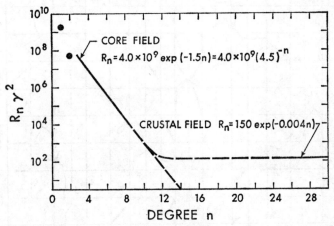

Fig. 4.19. Idealized surface power spectrum of the main geomagnetic field. (After Lowes, 1974a.)

Lowes also showed that coefficients of the main field for $n \gtrsim 9$ and secular variation coefficients for $n \gtrsim 6$ are not known at present with sufficient accuracy. The International Geomagnetic Reference Field (*IGRF*) is known to about 0·5 per cent at the Earth's surface but to only about 10 per cent at the surface of the core. Its time variation is known to about 20 per cent at the Earth's surface, and is very uncertain at the surface of the core.

It has been suggested (e.g. Hide, 1966) that the observed westward drift of the non-dipole part of the magnetic field is the surface expression of slow, free, hydromagnetic oscillations of the Earth's core in the presence of a dominant toroidal magnetic field. It is not possible however to settle the question of whether the observed westward drift is a wave phenomenon or corresponds to an average relative motion of the outer core. Hide's analysis was not completely rigorous and Stewartson (1967) later showed that in fact for such slow oscillations the drift is eastward, although for some higher modes a westward drift is also possible. Similar results were obtained by Malkus (1967) who found "nothing in the linear or non-linear aspects of the free wave solutions to suggest a preference for westward motion". Both Hide and Stewartson showed that hydromagnetic waves set up in a *thin* shell through the interaction between rotation and a toroidal magnetic field would drift eastward. Hide however maintained that there is a change in sign in the direction of drift in a *thick* shell—this was not substantiated in the later paper by Stewartson. More precisely Stewartson showed that the principal oscillations, i.e. those in which the radial component of velocity does not change sign along a radius vector drift eastward if $|m| > 1$ where m is the harmonic coefficient. When $|m| = 1$, the principal oscillations can drift westward, but this is unlikely to be the case. For higher modes of oscillation, waves which drift both eastward and westward can occur. This question has been discussed in more detail by Hide and Stewartson (1972) in a review paper on hydromagnetic oscillations of the Earth's core.

It is possible to make an order of magnitude estimate of the minimum power required to drive the dynamo. To maintain the electric current system in the core giving the present dipole field requires about $3 \times 10^{14} - 3 \times 10^{15}$ erg/sec: to maintain the toroidal field of about 100 gauss in the core requires about $10^{17} - 10^{18}$ erg/sec (Lowes, 1970). Even allowing for thermodynamic and mechanical inefficiencies, these values are small compared with the total power dissipated in the Earth. Braginskii (1964b), on the other hand, estimated rather higher power requirements—$\sim 4 \times 10^{19}$ erg/sec.

There is not a great variety of forces that can produce motions in the Earth's core. Most naturally occurring fluid motions are due ultimately to the action of gravity. The gravitational potential near the Earth consists of two components—that due to the Earth itself and that due to extraterrestrial sources (the sun and the moon). Consider first gravitational fields of extra terrestrial origin. The Earth's mantle undergoes complicated accelerations due to a number of causes such as the bodily tide of the Earth, tidal friction in the oceans, sudden changes in the rate of rotation of the Earth and precession and nutation. It can be shown (see e.g. Hide, 1956) that of these only precession could have any appreciable effect on motions in the Earth's core, and experiments by Malkus (1968) have indicated that precession may produce turbulent motion in the core and hence drive the dynamo. Malkus (1963) had earlier suggested that precessional torques might drive the Earth's dynamo, but unfortunately there are errors in his paper. A detailed theory of a dynamo in a precessing turbulent core is difficult and no full treatment has as yet been given. However, Rochester *et al.* (1975) have pointed out the mathematical and physical errors in Malkus' arguments and shown that precession fails by at least two orders of magnitude to satisfy the power requirements to drive the dynamo.

Precessional power input to the core (at the expense of the obliquity) can be estimated in terms of the dissipative part of the core-mantle coupling and the tiltover angle, i.e. the inclination of the core angular momentum vector to that of the mantle. Malkus failed to allow for the electrical conductivity contrast at the MCB and neglected the dependence of the coupling strength on the diurnal frequency of the precession-induced core flow relative to the mantle. Stacey (1973) had also argued that precessional torques can power the geodynamo. He also neglected the above dependence of the coupling strength and in addition estimated the tiltover angle by a non-rigorous kinematic argument. A detailed discussion of motions within the Earth's core has been given by Busse (1971). In an analysis of the flow in the core of a precessing Earth, he found that the angle between the axis of rotation of the core and that of the mantle in only 10^{-5} rad. The fact that the essentially toroidal velocity field obtained in his solution cannot by itself act as a dynamo, led him to the conclusion that precession cannot be the driving mechanism. The more detailed analysis of Rochester et al. (1975) indicates that the possibility of a precession driven dynamo is extremely remote.

Order of magnitude arguments on the feasibility of precession to drive the geodyanmo (e.g. Malkus, 1968) also depend quite critically on the value of the electrical conductivity σ of the core. There are two problems—the functional dependence on σ of various critical parameters such as the ohmic dissipation Q and the correct value of σ in the core. Rochester et al. (1975) have corrected errors in the literature and shown that Q in the core is proportional to σ^{-2}. For the value of σ, Malkus assigned an uncertainty of a factor of 3 in either direction to the value he assumed in his calculations (7×10^{-6} e.m.u.). In 1967 Stacey suggested that allowance for the effect of impurities in the iron in the Earth's core would increase its electrical resistivity by a factor of 10. However, in a later paper, Gardiner and Stacey (1971) withdrew this suggestion of such a large increase and preferred Bullard's (1949a) original estimate of 3×10^{-6} for the conductivity. They suggested that a plausible range of values of σ is $1-6 \times 10^{-6}$. In this respect it is interesting to note that Braginskii in 1964b estimated the conductivity of a core alloyed with 30 per cent Si to be just twice Bullard's figure. The normalization Gardiner and Stacey used to allow for metallic impurities in the iron is reasonable, as liquid silicon is itself metallic. However, their extrapolation of Bridgman's (1957) data on resistivities of solid iron alloys up to 100 kbar to liquid iron at Mbar pressures is not justified. Thus, although the conclusions of Gardiner and Stacey may well be right, some of their arguments in rebuttal of Stacey's (1967) earlier suggestion of a large increase in resistivity in the core appear to be incorrect (see Jacobs et al., 1972).

Jain and Evans (1972) have carried out a calculation of the resistivity of the Earth's core based on a model for the electrical transport properties of simple liquid metals proposed by Ziman (1961, 1971). Their estimate of the resistivity is between 1 and 2×10^5 e.m.u. which is in the range of plausible values according to Gardiner and Stacey. It is also of interest to note the results of the experimental work of Kawai and Mochizuki (1971) on the metallic states in three 3d transition metal oxides (including Fe_2O_3) under very high pressures. They found a very sudden drop in resistivity of from four to six orders of magnitude at a pressure corresponding to that in the liquid core of the Earth. Whether the resistivity of the Earth's core is of the same order of magnitude as that of pure iron at its melting temperature at atmospheric pressure will depend on the alloying material (see Section 5.5). If it is mainly silicon, the answer is probably yes since liquid silicon is a normal metal. If it is sulphur, the answer would be very different, since liquid sulphur is almost a perfect insulator. However, if the core material is in the form of the compound FeS, the resistivity can still be quite similar to that of pure iron since liquid FeS is a semi-conductor under normal temperature and pressure conditions and there may be a Mott transition to metal at very high pressures. Estimates of the conductivity obtained by Jain and Evans (1972) are probably the best to date and are in general argreement with experimental results obtained by Johnston and Strens (1973) on a molten Fe–Ni–S–C core mix.

The Earth's own gravitational field could generate core motions if there are density inhomogeneities. This could be brought about in two ways—by "sedimentation" or by thermal convection. Braginskii (1964a) suggested that during the crystallization of the inner core there was continuous formation of an excess of the light component (silicon), whose upward flotation is the basic factor responsible for convection in the core. The dissolution of silicon in iron is associated with a large decrease in volume and liberation of heat, since Fe and Si atoms in solution form stable FeSi complexes. Braginskii maintained that convection in the core must be non-thermal since the efficiency of any possible heat engine is so low that an impossibly large heat flux from the core would be required.

Again if material were continuously falling from the mantle into the core, fluid motions would result. Urey (1952) has suggested that the core has been growing at the expense of the mantle—iron in the mantle slowly and continuously "seeping" into the core—the heat which is released in the core by this process causing convection. In the past the mantle contained more free iron at which time precipitation from the mantle probably played a much greater role than now. The basic question is, however, how long did it take the Earth's core to form and when did this event take place? All evidence (see Section 2.4) indicates that in all probability the event was comparatively rapid and took place very early in, or simultaneously with, the formation of the Earth itself. It would thus seem that the processes which led to core formation have not continued throughout geologic time and thus could not be a major factor in the maintenance of motions in the Earth's core.

If there are heat sources in the core, thermal expansion will give rise to density inhomogeneities. In the absence of the inhibiting effects of viscosity, thermal conduction, rotation and magnetic fields, convection will occur if the heat supply exceeds that required to maintain the heat flow by conduction along the adiabatic gradient. To investigate thermal convection quantitatively the equations of thermodynamics must be considered in addition to the hydromagnetic equations. Again only order of magnitude estimates have so far been possible. There are two critical questions to be answered before the importance of thermal convection in the core can be decided—are there sufficient heat sources in the core and, if there are, how can this excess heat be disposed of since only a small fraction of the heat released would be converted into magnetic and kinetic energy?

We have no direct knowledge of any heat sources in the core. Order of magnitude calculations however indicate that a very small fraction of the radioactive content of acid igneous rocks would suffice. Bullard (1949a) suggested that there may be minute concentrations of uranium and thorium in the core and Elsasser (1946b) proposed that the heavy oxides of thorium and uranium sank to the centre of the Earth when it was formed. There are however a number of difficulties with these ideas (see e.g. Urey, 1952) and neutron activation analyses have failed to find any radioactivity in iron meteorites.

The possibility of the presence of sulphur in the core could indirectly have a profound effect on thermal convection, since it could also entail a concentration of potassium in the core. This question is discussed in some detail in Section 5.6. Hurley (1968) has estimated that from three quarters to seven eighths of the radio nuclide K^{40} on Earth may be located in the core, and thus a major source of heat, roughly 10^{20} erg/sec may exist within the core. This is more than sufficient to sustain convection in the core and an abundant supply of energy would be available for core motions to maintain the Earth's magnetic field, even if the efficiency of conversion is quite small.

Heat from the core may be disposed of if the thermal conductivity of the mantle is sufficiently high—as it may be if radiative transfer of heat is substantially increased at depth. The importance of radiative heat transfer in the mantle has however now been seriously questioned (Fukao, 1969; Pitt and Tozer, 1970a, b). If convection occurs in the deep mantle it would also greatly help in transferring heat from the core. On the other hand if the core is thermally insulated by the mantle, the temperature of the core as a whole would be raised—perhaps this is the reason for the outer core being fluid. If the additional heat is not excessive, it may keep the outer core approximately at its melting point.

Two recent papers have dealt with the problem of dynamo action by thermally driven convective motions of an electrically conducting fluid between two parallel plates. Busse (1973a) has shown that convective rolls alone cannot produce dynamo action, but if a shear is imposed along the rolls then the flow resembles one of G. O. Roberts' periodic dynamos. Although Busse's dynamo is unrealistic in its geometry, boundary conditions and large magnetic Prandtl number, it represents a substantial step forward from the kinematic dynamo. The problem lends itself to numerical treatment, and numerical solutions may provide information about the more complicated properties of hydromagnetic dynamos.

Childress and Soward (1972) have examined the more realistic problem of a rotating fluid between two plates, considering two interacting convecting rolls. Their work is particularly interesting since the equations are similar to those for coupled disk dynamos, which exhibit reversals (see Moffatt, 1973). Details of their work have not as yet been published.

Mullan (1973) has suggested that seismic waves may be sufficient to produce the motions necessary to drive the geomagnetic dynamo. He estimated that if all the earthquake energy (10^{26}–10^{27} erg/yr) were transferred to the whole liquid core as kinetic energy and if this kinetic energy were entirely dissipated in one year, the resulting motions would have an energy density of 1–10 erg/cm^3. He further estimated that the lifetime of motions against viscous damping is ~ 100 yr if the velocity variations have length scales of the order of 500 km or more and the kinematic viscosity $\leqslant 10^6$ cm^2/sec. Thus the kinetic energy available at any time may be derived from the previous 100 yr seismic activity. Mullan suggested that energy densities in the range 10^2–10^3 erg/cm^3 may be in equipartition with fields of 50–105 guass.

Not all the energy from earthquakes will be transmitted as kinetic energy in the core (Won and Kuo (1973) estimated that only 10^{-4} will reach the inner core). Mullan estimated however that even if only 10^{-4} of the seismic energy is converted into kinetic energy in the outer 100 km of the liquid core, this would be sufficient to achieve equipartition with a 5 gauss field at the MCB. Mullan also drew attention to the fact that most deep earthquakes occur in the circum-Pacific seismic belt and in the Pacific Ocean where both the non-dipole field and secular variation are anomalously low (see Section 4.1). He speculated whether the seismic excitation of core motions beneath the Pacific may be sufficiently intense to cause a "semi-permanent centre of magnetic action" with little drift permitted.

4.8. Variations in the Length of the Day

As a result of the attraction of the sun and moon on the Earth's equatorial bulge and of movements of mass within the Earth, the angular velocity of the

Earth is not constant—there are in fact fluctuations, not only in the rate of spin (i.e. changes in the length of the day), but also in the direction of the axis of rotation i.e. the Earth "wobbles". There are a number of peaks in the frequency spectrum of the Earth's rotation, covering a very long time scale— these peaks are believed to arise from different causes. Three distinct components have been recognized—a steady increase in the length of the day by about 2×10^{-3} sec a century, seasonal fluctuations of about 10^{-3} sec, and less regular variations up to about 5×10^{-3} sec having time scales of the order of years (the so-called decade fluctuations).

The seasonal variations in the length of the day are chiefly the result of torques on the mantle exerted by oceanic currents and atmospheric winds. The rapid irregular variations in the length of the day over a decade however cannot be explained by surface phenomena. No transport of mass at the surface could alter the Earth's moment of inertia by a sufficient amount to account for such large changes as are observed. It has been suggested that these "decade" fluctuations are caused by the transfer of angular momentum between the Earth's solid mantle and liquid core. This in turn implies some form of core–mantle coupling. There are a number of ways in which angular momentum could be transferred between the core and the mantle the principal possible mechanisms being inertial, electromagnetic and topographic coupling. Inertial coupling could arise from hydrodynamic pressure forces acting over the ellipsoidal MCB when internal flow is induced in the liquid core by a shift in the earth's rotation axis (Toomre, 1966). Electromagnetic coupling could arise from leakage of the secular variation into the electrically conducting lower mantle (Rochester, 1960, 1968). It is difficult, however, to make quantitative estimates of the horizontal stresses at the MCB. Calculations indicate that neither viscous coupling nor electromagnetic coupling is really adequate to account for the decade fluctuations (see e.g. Rochester, 1970, 1973). More recently Hide (1969) has suggested the possibility of topographic coupling as a result of irregular features (bumps) at the MCB (see Section 4.10). It would be interesting to try and detect any small departures in uniformity in the rate of rotation of other planets such as Mars and Venus and of planetary satellites with rigid surfaces, since any such departures may yield information on a possible liquid convecting core and planetary magnetic fields.

Changes over geologic time are predominantly a constant deceleration as a result of tidal friction. It is difficult to estimate the rate at which this deceleration has taken place since our knowledge of palaeogeography is scant and the result depends critically on a few shallow seas. Urey (1952) has suggested, on the other hand, that, because of differentiation of the materials of the Earth and the growth of the core, the moment of inertia of the Earth about its axis of rotation may have been reduced and as a result the length of the day decreased from about 30 to 24 hr. The changes caused by a growing core are considerably smaller (and of opposite sign) to those due to tidal friction (Runcorn, 1964, 1970a).

In both the atmosphere and the oceans the role of rotation is of fundamental importance. This is also true—with some qualifications—for the Earth's fluid core. With regard to rotational effects we must distinguish between the core and oceanic–atmospheric layers in the following manner. The core should only be divided up into thin layers if it is stably stratified (see Section 3.5). In such a case dynamo action, though still possible, would be significantly constrained to a predominantly two dimensional motion in concentric spherical shells. If, on the other hand, the core is not stratified, any perturbation of the otherwise steady rotation would lead to a predominantly two dimensional motion in planes perpendicular to the rotation axis as predicted by the Proudman-Taylor theorem. In both cases these statements must be modified to include the effects of the Lorentz force.

In certain special cases the effects of the Lorentz force can be included without too much difficulty. For example, we can consider the possibility of interpreting the westward drift of the non-dipole field as the propagation of hydromagnetic waves in the fluid core around the axis of rotation in the presence of a dominant toroidal magnetic field (Hide, 1966). If the fluid is *not* stably stratified the Lorentz force enters into the equations of motion for hydromagnetic wave solutions in the same manner as the Coriolis force. In fact the flow in this case is described by the Poincaré equation—the same equation that describes the flow in the absence of a magnetic field. But even in this rather simple example there is an added difficulty. We are looking for solutions to a hyperbolic differential equation with certain specified boundary conditions. This is an ill-posed mathematical problem and there are serious difficulties in attempting to find a solution. A major source of the difficulty is the presence of an inner boundary to the fluid viz. the solid inner core.

Yukutake (1973a) has examined in some detail his earlier (1972) suggestion that fluctuations in the Earth's rate of rotation are related to changes in the dipole moment. He investigated the possibility of such a relationship for three different time periods, 8000, 400 and 65 yr, using archeomagnetic data, observations of the variation of the moon's longitude and recent observatory data. He found that such a relationship does exist, but that it is highly dependent on the period, the magnitude of the change increasing as the period decreases. He was able to account for this dependence of the excitation of the change in rotation rate upon period by electromagnetic coupling between the mantle and core if the conductivity of the lower mantle is as large as 10^{-8} e.m.u. and if the change in the dipole field originates near the surface of the core.

From ancient observations of eclipses, the Earth's rotation is known to have been *accelerated* during the past few thousand years in addition to the steady retardation due to tidal friction. Archeomagnetic studies have shown that the dipole field has also been changing, approximately periodically, with a large amplitude ($\simeq 50$ per cent of the present dipole moment) and with a period $\simeq 8000$ yr (Bucha, 1967, 1970; Cox, 1968; Kitazawa, 1970). Since the last maximum of the dipole moment (sometime between 0 and A.D. 500), the electromagnetic coupling between the mantle and core has been diminishing and acceleration of the Earth's rotation is thus to be expected during the past 2000 yr. Observation confirms the theoretical prediction (Yukutake, 1972) of a phase difference of about π between rotation and dipole moment change, for periods $\simeq 8000$ yr.

Yukutake (1971) has also shown that the magnitude of the gauss coefficient g_1^0 increased during the seventeenth and eighteenth centuries and then began to

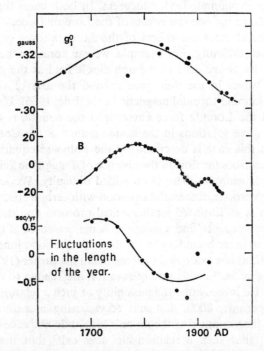

Fig. 4.20. Fluctuations in various features of the geomagnetic field and the Earth's rotation. From the top, fluctuations in the gauss coefficient $g_1{}^\circ$ (the dipole term), the moon's longitude (curve B), and the length of the year. (After Yukutake, 1973a.)

decrease since the early nineteenth century—this variation being superposed on the general trend discussed above (the gradual decrease since about 2000 yr ago). During this period there was also a large fluctuation in the observed longitude of the moon which has been ascribed to a change in the Earth's rate of rotation (see Fig. 4.20). The curve showing fluctuations in the length of the year leads the dipole curve by about $\pi/2$ for these periods ($\simeq 400$ yr).

The non-tidal variations in the length of the day must conserve the total angular momentum of the Earth: for variations of the order of a decade, the core of the Earth appears to be the only reasonable source. Yukutake (1973a, see Fig. 4.21) found a variation in the dipole field in phase with length of the year fluctuations ($\simeq 65$ yr period). Variations in the westward drift of the

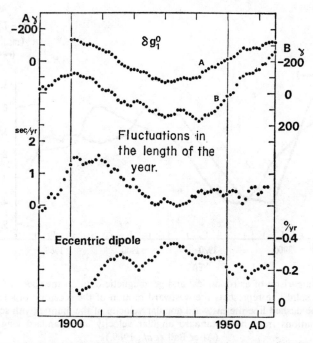

Fig. 4.21. Comparison of fluctuations in the length of the year with those of various features of the geomagnetic field. From the top, variations in the dipole term (δg_1°)—curve A from an analysis of 21 observatories, curve B from an analysis of 6 observatories—fluctuations in the length of the year and the drift rate of the eccentric dipole. (After Yukutake, 1973a.)

secular variation are however very different from those obtained earlier by Vestine (1953). Figure 4.22 (by Vestine and his co-workers) shows a comparison between the motion of the eccentric dipole and the deviation in the length of the day. Vestine and Kahle (1968) have interpreted these curves as showing that the angular velocity of the eccentric dipole is related to changes in the angular momentum of the outer $\simeq 200$ km of the core. Figure 4.22 also shows that there is an apparent phase lag between variations in the rate of rotation of the mantle and in the velocity of the eccentric dipole. Ball *et al.* (1969) suggested that this is the effect of diffusion of the magnetic signal through the conducting mantle. Yukutake (1973b) has questioned, however, whether the movement of the eccentric dipole really represents that of the geomagnetic field as a whole. He showed that the westward movement of the eccentric dipole over the last 150 yr is determined almost entirely by the westward drift arising from only one term ($n=2$, $m=1$) of the geomagnetic potential.

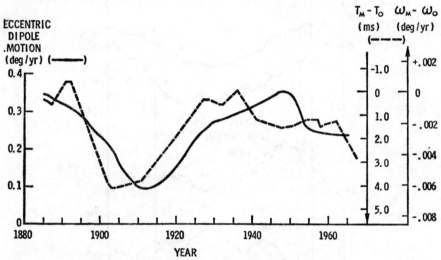

Fig. 4.22. Comparison of astronomical and geomagnetic data on the speed of the Earth's rotation. The solid line represents the westward motion of the eccentric dipole relative to the mantle, the dashed line the eastward angular velocity of the mantle (with scales appropriate to deviations from both standard angular velocity and standard length of day). (After Ball *et al.*, 1969.)

It is generally assumed that the westward drift of the non-dipole part of the Earth's magnetic field is a rotation about the geographical axis. Malin and Saunders (1973) have recently investigated the possibility that the secular change in the magnetic field might be better represented by rotation about another axis. They estimated pole positions and rotation rates which would give maximum correlation between magnetic field models for different epochs. They found that the pole of rotation has apparently moved from near Novaya Zemlya in 1945 to near the north magnetic pole in Canada in 1965 at a fairly steady rate (see Fig. 4.23). During this time, the drift rate remained near 0·18°/yr, approximately westwards, but showed a maximum amplitude when the rotation pole was near the geographical pole.

Malin and Saunders concluded that in general the pole of rotation differs significantly from the geographical pole—had the true pole of rotation coincided with the geographical pole, it would be expected that the points in Fig. 4.23 would be randomly distributed about the geographical pole. If the movements of the field reflect corresponding movements of the outer layers of the core, their results imply that the absolute rotation of the outer core is about an axis slightly different from that of the mantle, the rotation poles of the core and the mantle being on opposite sides of the total angular momentum pole (the direction of which is fixed in space).

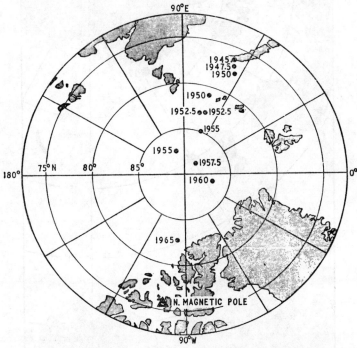

Fig. 4.23. Position of the pole of rotation of the non-dipole part of the Earth's magnetic field. (After Malin and Saunders, 1973.)

Lowes (1974b) has criticized the work of Malin and Saunders on a number of grounds. Since it is not possible to estimate satisfactorily the standard errors of the pole positions calculated by Malin and Saunders, they (1974) repeated their calculations for epochs earlier than 1940. The results agreed with their predictions viz. that the optimum pole positions would lie near the magnetic pole–Novaya Zemlya line and that their distance from the geographical axis would increase for earlier epochs (see Fig. 4.24). Malin and

Saunders also predicted that the rotation rate would be smaller for earlier epochs. This was also confirmed—actually the rotation rate drops to zero around 1930 and for earlier epochs has the opposite sign.

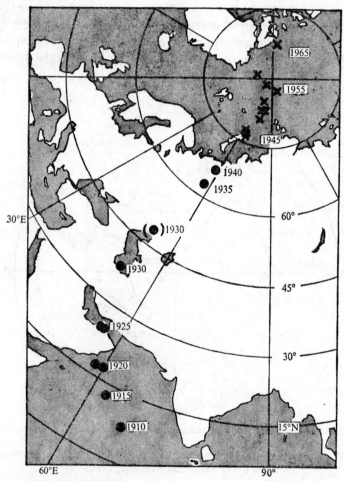

Fig. 4.24. Position of the optimum pole of rotation of the non-dipole part of the geomagnetic field. (After Malin and Saunders, 1974.)

4.9. Wobble of the Earth's Axis of Rotation

In addition to fluctuations in the Earth's rate of spin, there are also variations in the direction of the axis of rotation i.e. a "wobble". In 1891, S. C. Chandler isolated a component with a 14-month period (since called the Chandler wobble). For a rigid Earth it can be shown that the period of its free nutation is about 10 months—the effect of the Earth's deformation is thus to increase the period approximately 40 per cent. Observations show that the Chandler wobble is damped so that it must be constantly re-excited. Rochester and Smylie (1965) showed that electromagnetic core–mantle coupling fails by four orders of magnitude to provide the necessary damping.

The excitation mechanism of the Chandler wobble has been the subject of much debate. Changes in the annual variation of the mass distribution of the atmosphere fail by at least one order of magnitude (Munk and Hassan, 1961). A possible connection with seismic activity has often been proposed. Estimates of the contribution of earthquakes to the excitation of the Chandler wobble have until recently been several orders of magnitude too small, mainly because the displacement fields of even the largest earthquakes were thought to extend no more than a few hundred km from the focus. Following the work of Press (1965), which indicated that for great earthquakes a measureable displacement field may extend several thousand km from the epicentre, Mansinha and Smylie (1967, 1968) estimated, using dislocation theory, the changes in the products of inertia of the Earth arising from several large faults associated with major earthquakes. They found that the cumulative effect (based on earthquake statistics) could account for both the excitation of the Chandler wobble and a slow secular shift of the mean pole of rotation. However, Mansinha and Smylie (1970) and Dahlen (1971) disagree on whether the cumulative effect of all earthquakes is enough to sustain the Chandler wobble. Independent formulations of the theory for a model of a self-gravitating Earth with a liquid core and realistic distributions of density and elastic properties in the mantle have been given by Smylie and Manshinha (1971), Dahlen (1971, 1973) and Israel et al. (1973). Their theoretical treatments differ in detail and have given rise to some controversy over the physical principles governing static deformation of the liquid core. However, the effect of different boundary conditions at the MCB is likely to be small, and the authors are in general agreement that a major earthquake can produce a polar shift of the order of $0''.1$.

Rochester and Smylie (1965) have shown that electromagnetic coupling at the MCB is completely inadequate to excite the Chandler wobble. On the other hand Runcorn (1970b) believes that high-frequency components of the secular variation at the MCB are the core equivalent of sunspots and can supply an impulsive torque to the mantle that can transfer angular momentum rapidly enough to sustain the Chandler wobble. Earthquakes leave the instantaneous rotation pole unchanged but shift the axis of figure, so that the pole path experiences a discontinuous change in direction. Impulsive torques, on the other hand, leave the axis of figure unchanged and shift the rotation pole, so that the radius of the pole path is changed discontinuously. The details of electromagnetic coupling on such a short time scale have not been fully worked out, partly because high-frequency components of the secular variation are screened from observation by the electrical conductivity of the lower mantle that provides the coupling. Kakuta (1965) has concluded, however, that magnetohydrodynamic oscillations in the core could not excite a detectable wobble.

Longer period wobbles have also been suggested. Markowitz (1970) has adduced empirical evidence for a 24-yr period wobble, which Busse (1970) has suggested may represent the response of the mantle to a wobble of the solid inner core inertially coupled to the mantle via the liquid outer core. Rykhlova (1969), using a longer, but less homogeneous record, has found evidence for a 40-yr period. McCarthy (1972) also found, from latitude observations at Washington, a "period" somewhat longer than Markowitz's. If it is real, the phenomenon may well be the only observable manifestation, in the entire spectrum of changes in the Earth's rotation, of the presence of the solid inner core.

4.10. Topography of the Core-Mantle Boundary

Hide (1969) has suggested that the MCB may not be smooth*. Analysis of travel-times of compressional waves reflected at this boundary shows that any topographic feature there cannot exceed a few km in height—this being the limit of resolution of present-day seismic techniques. Hide suggested that irregular features at the MCB might provide "topographic" coupling between the core and mantle and so account for the decade fluctuations in the length of the day (neither viscous nor electromagnetic coupling seems adequate—see Section 4.8). Topographic "bumps" on the MCB may also be the cause of horizontal density variations responsible for regional gravity anomalies. It can readily be shown (e.g. Hide and Horai, 1968) that, because of the density contrast at the MCB, bumps with horizontal dimensions up to thousands of km and a km or so in height would make a significant (although not dominant) contribution to the observed distortion of the gravitational field at the Earth's surface. Hide (1967, 1969, 1970) also suggested that bumps on the MCB might affect the flow pattern in the core and thus influence the detailed configuration of the geomagnetic field and its time variations. While the liquid core of the Earth is the only likely location of electric currents responsible for the main geomagnetic field, it is the most unlikely place to find density variations of sufficient magnitude to cause the observed distortions of the

---

* The suggestion that the MCB may become "warped" as a result of stresses developed by currents in the outer part of the core was suggested earlier by Garland (1957).

gravitational field—these must arise largely in the mantle. Thus any correlation between gravity and magnetic anomalies should reflect processes at the MCB. Hide and Malin (1970) argued that if both gravity and magnetic anomalies are the result of the same topographic features, it should be possible to find a statistically significant correlation between them. In fact they found, for spherical harmonic coefficients up to degree 4, a correlation coefficient of 0·84 between large scale features of the Earth's non-dipole magnetic field (for

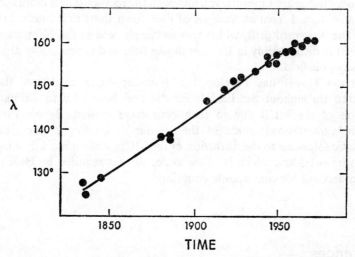

Fig. 4.25. The variation with time of the eastward displacement in longitude λ between the Earth's magnetic and gravitational fields. (After Hide and Malin, 1970.)

epoch 1965) and the gravitational field, provided the magnetic field is displaced 160° eastward in longitude λ. Hide and Malin also showed that λ has increased linearly with time since 1835 (see Fig. 4.25), the date of the earliest reliable spherical harmonic analysis (by Gauss) of the geomagnetic field. Their result is

$$\lambda = (126 \cdot 2 \pm 0 \cdot 2)° + (0 \cdot 273 \pm 0 \cdot 005)\,(t - 1835 \pm 10)° \qquad (4.14)$$

where t is the epoch (yr A.D.). This dependence of λ on t is associated with the westward drift of the geomagnetic field. There has been some controversy over the above correlation, particularly concerning the statistical procedures used (see e.g. Khan, 1971; Lowes, 1971, and reply by Hide and Malin, 1971). It must be stressed that even if the correlation does exist (as seems most likely) it does not by itself prove the existence of bumps on the MCB. It is possible that quite small temperature variations over the MCB could, through their effects on core motions, produce measurable distortions of the geomagnetic field. If these temperature variations in turn reflect the density structure of the lower mantle, then there would be a correlation between gravity and geomagnetic anomalies.

Baranova *et al.* (1973) have extended Hide and Malin's computation back to 1600 using spherical harmonic analyses of the geomagnetic field for 6 earlier epochs (1600, 1650 . . . 1850), Izmiran model D–1 (Ben'kova *et al.*, 1972). They found that a relationship between gravity and magnetic fields existed over the entire time interval, although the relationship is more complex than that obtained from an analysis of data for a short time interval (~ 100 years). The westward drift, which was so clearly seen in the $\lambda(t)$ pattern for 1829–1950 is distinct only in the non-dipole field and is practically absent in the quadrupole field.

Robinson (1974) has developed a boundary-layer model of thermal convection throughout the Earth's mantle and been able to estimate the distortion of the MCB due to such convective motion. By equating the additional gravitational force of the heavier descending plume with the hydrostatic force due to the distortion of the MCB, he estimated the displacement to be $\sim 1 \cdot 5$ km, which is of the order of that required by Hide (1969, 1970) to account for core-mantle coupling.

References

Acheson, D. J. and Hide, R. (1973). Hydromagnetics of rotating fluids. *Rep. Prog. Phys.* (UK) **36**, 159.

Allan, D. W. (1958). Reversals of the Earth's magnetic field. *Nature* **182**, 469.

Allan, D. W. (1962). On the behaviour of systems of coupled dynamos, *Proc. Camb. phil. Soc.* **58**, 671.

Backus, G. E. (1957). The axi-symmetric self-excited fluid dynamo. *Astrophys. J.* **125**, 500.

Backus, G. E. (1958). A class of self-sustaining dissipative spherical dynamos. *Ann. Phys.* **4**, 372.

Backus, G. E. (1968). Kinematics of geomagnetic secular variation in a perfectly conducting core. *Phil. Trans. R. Soc.* A **263**, 239.

Backus, G. E. and Chandrasekhar, S. (1956). On Cowling's theorem on the impossibility of self-maintained axi-symmetric homogeneous dynamos. *Proc. natn. Acad. Sci. U.S.A.* **42**, 105.

Ball, R. H., Kahle, A. B. and Vestine, E. H. (1969). Determination of surface motions of the Earth's core. *J. geophys. Res.* **74**, 3659.

Baranova, T. N., Ben'kova, N. P. and Freyzon, A. A. (1973). Correlation of geomagnetic gravity fields. *Geomag. Aeron.* **XIII**, 644.

Barbetti, M. and McElhinny, M. (1972). Evidence of a geomagnetic excursion 30,000 yr BP. *Nature* **239**, 327.

Ben'kova, N. P., Cherevko, T. N. and Adam, N. V. (1972). Izmiran, Preprint No. 35, 1972.

Blakely, R. J. and Cox, A. (1972). Detection of short geomagnetic polarity intervals from vector magnetic profiles. *Trans. Am. geophys. Un.* **53**, 974.

Bochev, A. (1969). Two and three dipoles approximating the Earth's main magnetic field. *Pure appl. Geophys.* **74**, 29.

Bondi, H. and Gold, T. (1950). On the generation of magnetism by fluid motion. *Mon. Not. R. astr. Soc.* **110**, 607.

Booker, J. R. (1969). Geomagnetic data and core motions. *Proc. R. Soc.* A **309**, 27.

Braginskii, S. I. (1964a). Kinematic models of the Earth's hydromagnetic dynamo. *Geomag. Aeron.* **IV**, 572.

Braginskii, S. I. (1964b). Magnetohydrodynamics of the Earth's core. *Geomag. Aeron.* **IV**, 698.

Braginskii, S. I. (1965a). Self-excitation of a magnetic field during the motion of a highly conducting fluid. Translated in *Sov. Phys. JETP* **20**, 726.

Braginskii, S. I. (1965b). Theory of the hydromagnetic dynamo. Translated in *Sov. Phys. JETP* **20**, 1462.

Braginskii, S. I. (1967). Magnetic waves in the Earth's core. *Geomag. Aeron.* **VII**, 851.

Braginskii, S. I. (1970a). Torsional magnetohydrodynamic vibrations in the Earth's core and variations in day length. *Geomag. Aeron.* **X**, 1.

Braginskii, S. I. (1970b). Oscillation spectrum of the hydromagnetic dynamo of the Earth. *Geomag. Aeron.* **X**, 172.

Braginskii, S. I. (1971). Origin of the geomagnetic field and its secular change. Trans. XV Gen. Assem. IUGG, Moscow (IAGA Bull. No. 31, 41).

Braginskii, S. I. (1972). Analytical description of the geomagnetic field of past epochs and determination of the spectrum of magnetic waves in the core of the Earth. *Geomag. Aeron.* **XII**, 947.

Bridgman, P. W. (1957). *Proc. Am. Acad. Arts Sci.* **84**, 179.

Brunhes, B. (1906). Recherches sur le direction d'aimantation des roches volcaniques. *J. Phys.* **5**, 705.

Bucha, V. (1967). Archaeomagnetic and palaeomagnetic study of the magnetic field of the Earth in the past 600,000 years. *Nature* **213**, 1005.

Bucha, V. (1970). Changes in the Earth's magnetic field during the archaeological past. *Comments Earth Sci. Geophys.* **1**, 20.

Bullard, E. C. (1949a). The magnetic field within the Earth. *Proc. R. Soc.* A **197**, 433.

Bullard, E. C. (1949b). Electromagnetic induction in a rotating sphere. *Proc. R. Soc.* A **199**, 413.

Bullard, E. C. (1955). The stability of a homopolar dynamo. *Proc. Camb. Phil. Soc.* **51**, 744.

Bullard, E. C. (1968). Reversals of the Earth's magnetic field. *Phil. Trans. R. Soc.* A **263**, 481.

Bullard, E. C., Freedman, C., Gellman, H. and Nixon, J. (1950). The westward drift of the Earth's magnetic field. *Phil. Trans. R. Soc.* A **243**, 67.

Bullard, E. C. and Gellman, H. (1954). Homogeneous dynamos and terrestrial magnetism. *Phil. Trans. R. Soc.* A **247**, 213.

Burg, J. P. (1967). Maximum entropy spectral analysis. Paper presented at the 37th meeting, Soc. Explor. Geophys., Oklahoma City, Okla, Oct. 31, 1967.

Burg, J. P. (1968). A new analysis technique for time series data. Paper presented at NATO Advanced Study Institute on Signal Processing, Aug. 1968, Enschede, Netherlands.

Busse, F. H. (1968). Steady flow in a precessing spheroidal shell. *J. Fluid Mech.* **33**, 739.

Busse, F. H. (1970). The dynamical coupling between inner core and mantle of the Earth and the 24-year libration of the pole. *In* "Earthquake Displacement Fields and the Rotation of the Earth" (Ed. L. Mansinha, D. E. Smylie and A. E. Beck), D. Reidel Publ. Co., Holland.

Busse, F. H. (1971). Bewegungen im kern der Erde. *Z. Geophys.* **37** (2), 153.

Busse, F. H. (1973a). Generation of magnetic fields by convection. *J. Fluid Mech.* **57**, 529.

Busse, F. H. (1973b). Private communication.

Childress, S. (1967). Construction of steady-state hydromagnetic dynamos. II: The spherical conductor. Courant Inst. Math. Sci. Rept. AFOSR-67-0976.

Childress, S. (1968). Théorie magnetohydrodynamique de l'effect dynamo. Rept. Dept. Mécanique, Fac. Sci. Paris. (Lectures delivered at l'Institut Henri Poincaré, Jan.–March, 1968).

Childress, S. (1970). New solutions of the kinematic dynamo problem. *J. math. Phys.* **11**, 3063.

Childress, S. and Soward, A. M. (1972). Convection-driven hydromagnetic dynamo. *Phys. Rev. Lett.* **29**, 837.

Cook, A. E. and Roberts, P. H. (1970). The Rikitake two-disk dynamo system. *Proc. Camb. Phil. Soc.* **68**, 547.

Cowling, T. G. (1934). The magnetic field of sunspots. *Mon. Not. R. astr. Soc.* **94**, 39.

Cowling, T. G. (1957). The dynamo maintenance of steady magnetic fields. *Q. Jl Mech. appl. Math.* **10**, 129.

Cowling, T. G. (1968). The axisymmetric dynamo. *Mon. Not. R. astr. Soc.* **140**, 547.

Cox, A. (1968). Length of geomagnetic polarity intervals. *J. geophys. Res.* **73**, 3247.

Cox, A. (1969). Geomagnetic reversals. *Science* **163**, 237.

Cox, A. (1972). Geomagnetic reversals, characteristic time constants, and stochastic processes. Int. Conf. Core-Mantle Interface. *Trans. Am. geophys. Un.* **53**, 613.

Cox, A. and Doell, R. R. (1964). Long period variations of the geomagnetic field. *Bull. seism. Soc. Am.* **54**, 2243.

Cox, A., Doell, R. R. and Dalrymple, G. B. (1964). Reversals of the Earth's magnetic field. *Science* **144**, 1537.

Crain, I. K. and Crain, P. L. (1970). New stochastic model for geomagnetic reversals. *Nature* **228**, 39.

Crain, I. K., Crain, P. L. and Plaut, M. G. (1969). Long period Fourier spectrum of geomagnetic reversals. *Nature* **223**, 283.

Creer, K. M. (1972). The behaviour of the paleo-geomagnetic field during reversals. Int. Conf. Core-Mantle Interface. *Trans. Am. geophys. Un.* **53**, 614.

Creer, K. M. and Ispir, Y. (1970). An interpretation of the behaviour of the geomagnetic field during polarity transitions. *Phys. Earth Planet. Int.* **2**, 283.

Creer, K. M., Thompson, R., Molyneux, L. and MacKereth, F. J. (1972). Geomagnetic secular variation recorded in the stable magnetic remanence of Recent sediments. *Earth Planet. Sci. Letters* **14**, 115.

Currie, R. G. (1966). The geomagnetic spectrum—40 days to 5·5 years. *J. geophys. Res.* **71**, 4579.

Currie, R. G. (1968). Geomagnetic spectrum of internal origin and lower mantle conductivity. *J. geophys. Res.* **73**, 2779.

Currie, R. G. (1973a). Geomagnetic time spectra—2 to 70 years. *Astrophys. Space Sci.* **21**, 425.

Currie, R. G. (1973b). The ~ 60 year spectral line in length of day fluctuations. *S. Afr. J. Sci.* **69**, 180.

Currie, R. G. (1973c). Pacific region anomaly in the geomagnetic spectrum at ~ 60 years. *S. Afr. J. Sci.* **69**, 379.

Dagley, P. and Lawley, E. (1974). Palaeomagnetic evidence for the transitional behaviour of the geomagnetic field. *Geophys. J.* **36**, 577.

Dagley, P., Wilson, R. L., Ade-Hall, J. M., Walker, G. P. L., Haggerty, S. E., Sigurgeirsson, T., Watkins, N. D., Smith, P. J., Edwards, J. and Grasty, R. L. (1967). Geomagnetic polarity zones for Icelandic lavas. *Nature* **216**, 25.

Dahlen, F. A. (1971). The excitation of the Chandler wobble by earthquakes. *Geophys. J.* **25**, 157.

Dahlen, F. A. (1973). A correction to the excitation of the Chandler wobble by earthquakes. *Geophys. J.* **32**, 203.

Doell, R. and Dalrymple, G. B. (1966). Geomagnetic polarity epochs: a new polarity event and the age of the Brunhes/Matuyama boundary. *Science* **152**, 1060.

Doell, R. R. and Cox, A. (1971). Pacific geomagnetic secular variation. *Science* **171**, 248.

Doell, R. R. and Cox, A. (1972). The Pacific geomagnetic secular variation anomaly and the question of lateral uniformity in the lower mantle. *In* "The Nature of the Solid Earth", (Ed. E. C. Robertson), McGraw-Hill, New York.

Dunn, J. R., Fuller, M., Ito, H. and Schmidt, V. A. (1971). Palaeomagnetic study of a reversal of the Earth's magnetic field. *Science* **172**, 840.

Elsasser, W. M. (1946a). Induction effects in terrestrial magnetism. Part I: Theory. *Phys. Rev.* **69**, 106.

Elsasser, W. M. (1946b). Induction effects in terrestrial magnetism. Part II: The secular variation. *Phys. Rev.* **70**, 202.

Elsasser, W. M. (1947). Induction effects in terrestrial magnetism. Part III: Electric modes. *Phys. Rev.* **72**, 821.

Elsasser, W. M. (1950). The Earth's interior and geomagnetism. *Rev. mod. Phys.* **22**, 1.

Elsasser, W. M. (1954). Dimensional values in magnetohydrodynamics. *Phys. Rev.* **95**, 1.

Frazer, M. C. (1973a). A search for axisymmetric dynamos. *Phys. Earth Planet. Int.* **7**, 111.

Frazer, M. C. (1973b). The dynamo problem and the geomagnetic field. *Contemp. Phys.* **14**, 213.

Fukao, Y. (1969). On the radiative heat transfer and the thermal conductivity in the upper mantle. *Bull. Earthq. Res. Inst.* **47**, 549.

Gailitis, A. (1970). The self-excitation of a magnetic field in a pair of vortex rings. *Magnitnaya Gidrodinamika* **6**, 19.

Gans, R. F. (1971). On hydromagnetic oscillations in a rotating cavity. *J. Fluid Mech.* **50**, 449.

Gardiner, R. B. and Stacey, F. D. (1971). Electrical resistivity of the core. *Phys. Earth Planet. Int.* **4**, 406.

Garland, G. D. (1957). The figure of the Earth's core and the non-dipole field. *J. geophys. Res.* **62**, 486.

Gibson, R. D. and Roberts, P. H. (1969). The Bullard–Gellman dynamo. *In* "The Application of Modern Physics to the Earth and Planetary Interiors" (Ed. S. K. Runcorn), Wiley, New York.

Gilliland, J. M. (1973). Meanfield electrodynamics and dynamo theories of planetary magnetic fields. Ph.D. Thesis, University of Alberta, 1973.

Gubbins, D. (1972). Kinematic dynamos and geomagnetism. *Nature Phys. Sci.* **238**, 119.

Gubbins, D. (1973). Numerical solutions of the kinematic dynamo problem. *Phil. Trans. R. Soc.* A **274**, 493.

Gubbins, D. (1974). Theories of the geomagnetic and solar dynamos. *Rev. Geophys. Space Phys.* **12**, 137.

Harrison, C. G. A. and Somayajulu, B. L. K. (1966). Behaviour of the Earth's magnetic field during a reversal. *Nature* **212**, 1193.

Heirtzler, J. R., Dickson, G. O., Herron, E. M., Pitman III, W. C. and LePichon, X. (1968). Marine magnetic anomalies, geomagnetic field reversals and motions of the ocean floor and continents. *J. geophys. Res.* **73**, 2119.

Helsley, C. E. (1972). Post Paleozoic magnetic reversals. *Trans. Am. geophys. Un.* **53**, 363.

Helsley, C. E. and Steiner, M. B. (1969). Evidence for long intervals of normal polarity during the Cretaceous period. *Earth Planet. Sci. Letters* **5**, 325.

Herzenberg, A. (1958). Geomagnetic dynamos. *Phil. Trans. R. Soc.* A **250**, 543.

Hide, R. (1956). The hydrodynamics of the Earth's core. *In* "Physics and Chemistry of the Earth", Vol. I, 94. Pergamon Press, Oxford.

Hide, R. (1966). Free hydromagnetic oscillations of the Earth's core and the theory of the geomagnetic secular variation. *Phil. Trans. R. Soc.* A **259**, 615.

Hide, R. (1967). Motions of the Earth's core and mantle and variations of the main geomagnetic field. *Science* **157**, 55.

Hide, R. (1969). Interaction between the Earth's liquid core and solid mantle. *Nature* **222**, 1055.

Hide, R. (1970). On the Earth's core-mantle interface. *Q. Jl R. met. Soc.* **96**, 579.

Hide, R. and Horai, K.-I. (1968). On the topography of the core–mantle interface. *Phys. Earth Planet. Int.* **1**, 305.

Hide, R. and Malin, S. R. C. (1970). Novel correlations between global features of the Earth's gravitational and magnetic fields. *Nature* **225**, 605.

Hide, R. and Malin, S. R. C. (1971). Novel correlations between global features of the Earth's gravitational and magnetic fields: further statistical considerations. *Nature Phys. Sci.* **230**, 63.

Hide, R. and Stewartson, K. (1972). Hydromagnetic oscillations of the Earth's core. *Rev. Geophys. Space Phys.* **10**, 579.

Honkura, Y. and Rikitake, T. (1972). Core motion as inferred from drifting and standing non-dipole fields. *J. Geomagn. Geoelect.* **24**, 223.

Hurley, P. M. (1968). Corrections to: Absolute abundance and distribution of Rb, K and Sr in the Earth. *Geochim. cosmochim. Acta* **32**, 1025.

IAGA Bulletin No. 28, The World Magnetic Survey 1957–1969 (Ed. A. J. Zmuda), IUGG Publ. Off., Paris, 1971.

Israel, M., Ben-Menahem, A. and Singh, S. J. (1973). Residual deformation of real Earth models with application to the Chandler wobble. *Geophys. J.* **32**, 219.

Jacobs, J. A. (1963). The Earth's Core and Geomagnetism. Pergamon Press, Oxford.

Jacobs, J. A. (1971). Reversals of the Earth's magnetic field. *Nature* **230**, 574.

Jacobs, J. A., Chan, T. and Frazer, M. C. (1972). Precession and the Earth's magnetic field. *Nature* **235**, 24.

Jain, A. and Evans, R. (1972). Calculation of the electrical resistivity of liquid iron in the Earth's core. *Nature Phys. Sci.* **235**, 165.

James, R. W. (1968). An equation for estimating westward drift. *J. Geomagn. Geoelect.* **20**, 429.

James, R. W. (1970). Decomposition of geomagnetic secular variation into drifting and non-drifting components. *J. Geomagn. Geoelect.* **22**, 241.

James, R. W. (1971). More on secular variation. *Comments Earth Sci. Geophys.* **2**. 28.

Jin, Rong-Sheng (1973). Dipole moments of the Earth, 1905–1966. *Trans. Am. geophys. Un.* **54**, 235.

Johnston, M. J. S. and Strens, R. G. J. (1973). Electrical conductivity of molten Fe–Ni–S–C core mix. *Phys. Earth Planet. Int.* **7**, 217.

Kahle, A. B., Ball, R. H. and Vestine, E. H. (1967). Comparison of estimates of fluid motions at the surface of the Earth's core for various epochs. *J. geophys. Res.* **72**, 4917.

Kahle, A. B., Ball, R. H. and Cain, J. C. (1969). Prediction of geomagnetic secular change confirmed. *Nature* **240**, 165.

Kakuta, C. (1965). Magnetohydrodynamic oscillation within the fluid core and irregularities in the rotational motion of the Earth. *Publs int. Latit. Obs. Mizasawa* **5**, 17.

Kawai, N. and Hirooka, K. (1967). Wobbling motion of the geomagnetic dipole field in historic time during these 2000 years. *J. Geomagn. Geoelect.* **19**, 217.

Kawai, N. and Mochizuki, S. (1971). Metallic states in the three 3d transition metal oxides, Fe_2O_3, Cr_2O_3 and TiO_2 under static high pressure. *Phys. Letters* **36A**, 54.

Kawai, N., Nakajima, T., Hirooka, K. and Kobayashi, K. (1973). The oscillation of field in the Matuyama geomagnetic epoch and the fine structure of the geomagnetic transition. *In* "Rock Magnetism and Paleogeophysics" (Ed. M. Kono), **1**, 53, Tokyo.

Khan, M. A. (1971). Correlations between the Earth's gravitational and magnetic fields. *Nature Phys. Sci.* **230**, 57.

Kitazawa, K. (1970). Intensity of the geomagnetic field in Japan for the past 10,000 years. *J. geophys. Res.* **75**, 7403.

Kovacheva, M. (1969). Inclination of the Earth's magnetic field during the last 2000 years in Bulgaria. *J. Geomagn. Geoelect.* **21**, 573.

Lacoss, R. T. (1971). Data adaptive spectral analysis methods. *Geophys.* **36**, 661.

Larmor, J. (1919). How could a rotating body such as the sun become a magnet? *Rep. Br. Ass. Advmt. Sci.* **159**.

Larson, E. E., Watson, D. E. and Jennings, W. (1971). Regional comparison of a Miocene geomagnetic transition in Oregon and Nevada. *Earth Planet. Sci. Letters* **11**, 391.

Lee, W. H. K. and Uyeda, S. (1965). Review of heat flow data. *In* "Terrestrial Heat Flow", Geophysical Monograph No. 8, *Am. geophys. Un.*

Leorat, J. (1969). Origine des champs magnétiques terrestre et solaire—théorie des dynamos turbulentes. Thèse Fac. Sci. Paris, Doctorat 3e Cycle, 1969.

Lilley, F. E. M. (1970). On kinematic dynamos. *Proc. R. Soc.* A **316**, 153.

Lortz, D. (1968). Impossibility of steady dynamos with certain symmetries. *Phys. Fluids* **11**, 913.

Lowes, F. J. (1970). Possible evidence on core evolution from geomagnetic dynamo theories. *Phys. Earth Planet. Int.* **2**, 382.

Lowes, F. J. (1971). Significance of the correlation between spherical harmonic fields. *Nature Phys. Sci.* **230**, 61.

Lowes, F. J. (1972). The recent geomagnetic field and its time variation. Int. Conf. Core–Mantle Interface. *Trans. Am. geophys. Un.* **53**, 610.

Lowes, F. J. (1974a). Spatial power spectrum of the main geomagnetic field and extrapolation to the core. *Geophys. J.* **36**, 717.

Lowes, F. J. (1974b). Rotation of the geomagnetic field. *Nature* **248**, 402.

Lowes, F. J. and Wilkinson, I. (1963). Geomagnetic dynamo: A laboratory model. *Nature* **198**, 1158.

Malin, S. R. C. and Saunders, I. (1973). Rotation of the Earth's magnetic field. *Nature* **245**, 25.

Malin, S. R. C. and Saunders, I. (1974). Rotation of the geomagnetic field—Reply. *Nature* **248**, 403.

Malkus, W. V. R. (1963). Precessional torques as the cause of geomagnetism. *J. geophys. Res.* **68**, 2871.

Malkus, W. V. R. (1967). Hydromagnetic planetary waves. *J. Fluid Mech.* **28**, 793.

Malkus, W. V. R. (1968). Precession of the Earth as the cause of geomagnetism. *Science* **160**, 259.

Malkus, W. V. R. (1971). Motions in the fluid core. *In* "Mantle and Core in Planetary Physics" (Ed. J. Coulomb and M. Caputo), Proc. Internat. School of Physics "Enrico Fermi", Course L, Academic Press, New York.

Malkus, W. V. R. (1972). Reversing Bullard's dynamo. Int. Conf. Core–Mantle Interface. *Trans. Am. geophys. Un.* **53**, 617.

Mansinha, L. and Smylie, D. E. (1967). Effect of earthquakes on the Chandler wobble and the secular polar shift. *J. geophys. Res.* **72**, 4731.

Mansinha, L. and Smylie, D. E. (1968). Earthquakes and the Earth's wobble. *Science* **161**, 1127.

Mansinha, L. and Smylie, D. E. (1970). Seismic excitation of the Chandler wobble. *In* "Earthquake Displacement Fields and the Rotation of the Earth" (Ed. L. Mansinha, D. E. Smylie and A. E. Beck), D. Reidel Publ. Co., Holland.

Markowitz, W. (1970). Sudden changes in rotational acceleration of the Earth and secular motion of the pole. *In* "Earthquake Displacement Fields and the Rotation of the Earth" (Ed. L. Mansinha, D. E. Smylie and A. E. Beck), D. Reidel Publ. Co., Holland.

Márton, P. (1970). Secular variation of the geomagnetic virtual dipole field during the last 2000 years as inferred from the spherical harmonic analysis of the available archeomagnetic data. *Pure appl. Geophys.* **81**, 163.

McCarthy, D. D. (1972). Secular and non-polar variation of Washington latitude. *In* "Rotation of the Earth" (Ed. P. Melchior and S. Yumi), D. Reidel Publ. Co., Holland.

McDonald, K. L. and Gunst, R. H. (1968). Recent trends in the Earth's magnetic field. *J. geophys. Res.* **73**, 2057.

McElhinny, M. W. (1971). Geomagnetic reversals during the Phanerozoic. *Science* **172**, 157.

McElhinny, M. W. and Evans, M. E. (1968). An investigation of the strength of the geomagnetic field in the early Pre-cambrian. *Phys. Earth Planet. Int.* **1**, 485.

Moffatt, H. K. (1970). Turbulent dynamo action at low magnetic Reynolds number. *J. Fluid Mech.* **41**, 435.

Moffatt, H. K. (1972). An approach to a dynamic theory of dynamo action in a rotating conducting fluid. *J. Fluid Mech.* **53**, 385.

Moffatt, H. K. (1973). Report on the NATO Advanced Study Institute on magneto-hydrodynamic phenomena in rotating fluids. *J. Fluid Mech.* **57**, 625.

Mullan, D. J. (1973). Earthquake waves and the geomagnetic dynamo. *Science* **181**, 553.

Munk, W. H. and Hassan, E. S. M. (1961). Atmospheric excitation of the Earth's wobble. *Geophys. J.* **4**, 339.

Nagata, T. (1969). Length of geomagnetic polarity intervals. *J. Geomagn. Geoelect.* **21**, 701.

Namikawa, T. and Matsushita, S. (1970). Kinematic dynamo problem. *Geophys. J.* **19**, 395.

Néel, L. (1951). L'inversion de l'aimantation permanente des roches. *Annls Geophys.* **7**, 90.

Néel, L. (1955). Some theoretical aspects of rock magnetism. *Phil. Mag. Supp. Adv. Phys.* **4**, 191.

Nelson, J. H., Hurwitz, L. and Knapp, D. G. (1962). Magnetism of the Earth. Publ. 40–1, U.S. Dept. Comm. Coast Geod. Surv., Washington, 1962.

Opdyke, N. D. (1969). The Jaramillo event as detected in oceanic cores. *In* "The Application of Modern Physics to the Earth and Planetary Interiors" (Ed. S. K. Runcorn), Wiley-Interscience, New York.

Opdyke, N. D. (1972). Paleomagnetism of deep-sea cores. *Rev. Geophys. Space Phys.* **10**, 213.

Opdyke, N. D., Glass, B., Hays, J. D. and Foster, J. (1966). Paleomagnetic study of Antarctic deep-sea cores. *Science* **154**, 349.

Opdyke, N. D., Kent, D. V. and Lowrie, W. (1973). Details of magnetic polarity transitions recorded in a high-deposition rate deep-sea core. *Earth Planet. Sci. Letters* **20**, 315.

Orlov, V. P. (1965). The leading trends of the secular variation investigation. *J. Geomagn. Geoelect.* **17**, 277.

Parker, E. N. (1955). Hydromagnetic dynamo models. *Astrophys. J.* **122**, 293.

Parker, E. N. (1969). The occasional reversal of the geomagnetic field. *Astrophys. J.* **158**, 815.

Pitt, G. D. and Tozer, D. C. (1970a). Optical absorption measurements on natural and synthetic ferromagnesian minerals subjected to high pressures. *Phys. Earth Planet. Int.* **2**, 179.

Pitt, G. D. and Tozer, D. C. (1970b). Radiative heat transfer in dense media and its magnitude in olivines and some other ferromagnesian minerals under typical upper mantle conditions. *Phys. Earth Planet. Int.* **2**, 189.

Press, F. (1965). Displacements, strains and tilts at teleseismic distances. *J. geophys. Res.* **70**, 2395.

Pudovkin, I. M. and Valuyeva, G. Ye. (1967). Causes of the so-called westward drift of the geomagnetic field. *Geomag. Aeron.* **VII**, 754.

Pudovkin, I. M. and Valuyeva, G. Ye. (1972). Nature of the drift of the main eccentric geomagnetic dipole. *Geomag. Aeron.* **XII**, 453.

Reid, A. B. (1972). A palaeomagnetic study at 1800 million years in Canada. Ph.D. Thesis, University of Alberta, 1972.

Rikitake, T. (1958). Oscillations of a system of disk dynamos. *Proc. Camb. Phil. Soc.* **54**, 89.

Rikitake, T. (1966a). Electromagnetism and the Earth's Interior. Elsevier Publ. Co., Amsterdam.

Rikitake, T. (1966b). Westward drift of the equatorial component of the Earth's magnetic dipole. *J. Geomagn. Geoelect.* **18**, 383.

Rikitake, T. (1967). Non-dipole field and fluid motions in the Earth's core. *J. Geomagn. Geoelect.* **19**, 129.

Rikitake, T. and Hagiwara, Y. (1966). Non-steady state of a Herzenberg dynamo. *J. Geomagn. Geoelect.* **18**, 393.

Roberts, G. O. (1970). Spatially periodic dynamos. *Phil. Trans. R. Soc.* A **266**, 535.

Roberts, G. O. (1972). Dynamo action of fluid motions with two-dimensional periodicity. *Phil. Trans. R. Soc.* A **271**, 411.

Roberts, P. H. (1967a). "An Introduction to Magnetohydrodynamics." Longmans, London.

Roberts, P. H. (1967b). The dynamo problem. *Woods Hole Ocean. Inst. Rept.* 67–54, Vol. 1, 51 and 178.

Roberts, P. H. (1971a). Dynamo theory. *In* "Mathematical Problems in the Geophysical Sciences" (Ed. W. H. Reid), Lectures in Applied Mathematics, Vol. 14, *Am. math. Soc.*

Roberts, P. H. (1971b). Dynamo theory of geomagnetism. *In* "World Magnetic Survey 1957–1969" (Ed. A. J. Zmuda), IAGA Bull. 28, IUGG Publ. Off., Paris.

Roberts, P. H. (1972). Kinematic dynamo models. *Phil. Trans. R. Soc.* A **272**, 663.

Roberts, P. H. and Stix, M. (1971). The turbulent dynamo: a translation of a series of papers by F. Krause, K.-H. Rädler and M. Steenbeck, NCAR Tech. Note 1A–60, Boulder, Colo.

Roberts, P. H. and Soward, A. M. (1972). Magnetohydrodynamics of the Earth's core. *In* Ann. Rev. Fluid Mech., Vol. 4.

Robinson, J. L. (1974). A note on convection in the Earth's mantle. *Earth Planet. Sci. Letters* **21**, 190.

Rochester, M. G. (1960). Geomagnetic westward drift and irregularities in the Earth's rotation. *Phil. Trans. R. Soc.* A **252**, 531.

Rochester, M. G. (1968). Perturbations in the Earth's rotation and geomagnetic core–mantle coupling. *J. Geomagn. Geoelect.* **20**, 387.

Rochester, M. G. (1970). Core–mantle interactions: geophysical and astronomical consequences. *In* "Earthquake Displacement Fields and the Rotation of the Earth" (Ed. L. Mansinha, D. E. Smylie and A. E. Beck), D. Reidel Publ. Co., Holland.

Rochester, M. G. (1973). The Earth's Rotation, *Trans. Am. geophys. Un.* **54**, 769.

Rochester, M. G. and Smylie, D. E. (1965). Geomagnetic core-mantle coupling and the Chandler wobble. *Geophys. J.* **10**, 289.

Rochester, M. G., Jacobs, J. A., Smylie, D. E. and Chong, K. F. (1975). Can precession power the geomagnetic dynamo? *Geophys. J.* (In press.)

Runcorn, S. K. (1964). Changes in the Earth's moment of inertia. *Nature* **204**, 823.

Runcorn, S. K. (1970a). Palaeontological measurements of the changes in the rotation rates of Earth and moon and of the rate of retreat of the moon from the Earth. *In* "Palaeogeophysics" (Ed. S. K. Runcorn), Academic Press, London.

Runcorn, S. K. (1970b). A possible cause of the correlation between earthquakes and polar motions. *In* "Earthquake Displacement Fields and the Rotation of the Earth" (Ed. L. Mansinha, D. E. Smylie and A. E. Beck), D. Reidel Publ. Co., Holland.

Rykhlova, L. V. (1969). Evaluation of the Earth's free nutation parameters from 119 years of observations. *Sov. Astron.* **AJ 13**, 544.

Skiles, D. D. (1970). A method of inferring the direction of drift of the geomagnetic field from palaeomagnetic data. *J. Geomagn. Geoelect.* **22**, 441.

Smith, P. J. (1967). The intensity of the ancient geomagnetic field: a review and analysis. *Geophys. J.* **12**, 321.

Smylie, D. E. and Mansinha, L. (1971). The elasticity theory of dislocations in real earth models and changes in the rotation of the Earth. *Geophys. J.* **23**, 329.

Soward, A. M. (1972). A kinematic theory of large magnetic Reynolds number dynamos. *Phil. Trans. R. Soc.* A **272**, 431.

Stacey, F. D. (1967). Electrical resistivity of the Earth's core. *Earth Planet. Sci. Letters* **3**, 204.

Stacey, F. D. (1973). The coupling of the core to the precession of the Earth. *Geophys. J.* **33**, 47.

Steenbeck, M. and Krause, F. (1966). Erklarung stellarer und planetarer magnet-felder durch ein turbulenzbedingten dynamomechanismus. *Z. Naturforsch* **21a**, 1285 (see P. H. Roberts and M. Stix, 1971).

Steinhauser, P. and Vincenz, S. A. (1973). Equatorial paleopoles and behaviour of the dipole field during polarity transitions. *Earth Planet. Sci. Letters* **19**, 113.

Stewart, A. D. and Irving, E. (1973). Palaeomagnetism of Precambrian red beds from N.W. Scotland. *Trans. Am. geophys. Un.* **54**, 248.

Stewartson, K. (1967). Slow oscillations of fluid in a rotating cavity in the presence of a toroidal magnetic field. *Proc. R. Soc.* A **299**, 173.

Stewartson, K. (1971). Planetary waves. *In* "Mathematical Problems in the Geophysical Sciences" (Ed. W. H. Reid), Lectures in Applied Mathematics, Vol. 14, *Am. Math. Soc.*

Toomre, A. (1966). On the coupling of the Earth's core and mantle during the 26,000-year precession. *In* "The Earth–Moon System" (Ed. B. G. Marsden and A. G. W. Cameron), Plenum Press, New York.

Tverskoy, B. A. (1966). Theory of hydrodynamic self-excitation of regular magnetic fields. *Geomagn. Aeron.* **VI**, 7.

Ulrych, T. J. (1972). Maximum entropy power spectrum of truncated sinusoids. *J. geophys. Res.* **77**, 1396.

Urey, H. C. (1952). "The Planets: Their Origin and Development", Yale University Press.

Verosub, K. L. and Cox, A. (1971). Changes in the total magnetic energy external to the Earth's core. *J. Geomagn. Geoelect.* **23**, 235.

Vestine, E. H. (1953). On variations of the geomagnetic field, fluid motion, and the rate of the Earth's rotation. *J. geophys. Res.* **58**, 127.

Vestine, E. H. and Kahle, A. B. (1968). The westward drift and geomagnetic secular change. *Geophys. J.* **15**, 29.

Vestine, E. H., Laporte, L., Cooper, C., Lange, I. and Hendrix, W. C. (1947). "Description of the Earth's main magnetic field and its secular change, 1905–1945", Carnegie Inst. Wash. Publ. No. 578.

Vogt, P. R., Einwich, A. and Johnson, G. L. (1972). A preliminary Jurassic and Cretaceous reversal chronology from marine magnetic anomalies in the western North Atlantic. *Trans. Am. geophys. Un.* **53**, 363.

Watkins, N. D. (1967). Unstable components and palaeomagnetic evidence for a geomagnetic polarity transition. *J. Geomagn. Geoelect.* **19**, 63.

Watkins, N. D. (1969). Non-dipole behaviour during an upper Miocene geomagnetic polarity transition in Oregon. *Geophys. J.* **17**, 121.

Wilson, R. L., Dagley, P. and McCormack, A. G. (1972). Palaeomagnetic evidence about the source of the geomagnetic field. *Geophys. J.* **28**, 213.

Won, I. J. and Kuo, J. T. (1973). Oscillation of the Earth's inner core and its relation to the generation of geomagnetic field. *J. geophys. Res.* **78**, 905.

Yaskawa, K. (1973). Reversals in Brunhes normal polarity epoch. *In* "Rock Magnetism and Paleogeophysics" (Ed. M. Kono), 1, 44, Tokyo.

Yukutake, T. (1962). The westward drift of the magnetic field of the Earth. *Bull. Earthq. Res. Inst.* **40**, 1.

Yukutake, T. (1968a). The drift velocity of the geomagnetic secular variation. *J. Geomagn. Geoelect.* **20**, 403.

Yukutake, T. (1968b). Two methods of estimating the drift rate of the Earth's magnetic field. *J. Geomagn. Geoelect.* **20**, 427.

Yukutake, T. (1970). Geomagnetic secular variation. *Comments Earth Sci. Geophys.* **1**, 55.

Yukutake, T. (1971). Spherical harmonic analysis of the Earth's magnetic field for the 17th and 18th centuries. *J. Geomagn. Geoelect.* **23**, 39.

Yukutake, T. (1972). The effect of change in the geomagnetic dipole moment on the rate of the Earth's rotation. *J. Geomagn. Geoelect.* **24**, 19.

Yukutake, T. (1973a). Fluctuations in the Earth's rate of rotation related to changes in the geomagnetic dipole field. *J. Geomagn. Geoelect.* **25**, 195.

Yukutake, T. (1973b). The eccentric dipole, and inadequate representation of movement of the geomagnetic field as a whole. *J. Geomagn. Geoelect.* **25**, 231.

Yukutake, T. and Tachinaka, H. (1969). Separation of the Earth's magnetic field into the drifting and the standing parts. *Bull. Earthq. Res. Inst.* **47**, 65.

Ziman, J. M. (1961). A thoery of the electrical properties of liquid metals. 1: The monovalent metals. *Phil. Mag.* **6**, 1013.

Ziman, J. M. (1971). The calculation of Bloch functions. *Solid St. Phys.* **26**, 1.

5. The Constitution of the Core

5.1. Equations of State

By an equation of state we mean a relationship between the pressure, specific volume and temperature of a material. An equation of state cannot involve the history of the material and thus non-hydrostatic stresses are excluded—large non-hydrostatic stresses in general lead to irreversible (plastic) deformation. The pressures and temperatures in the deep interior of the Earth are on the one hand sufficiently high to make them difficult to reproduce in the laboratory, and on the other hand sufficiently low to make the quantum-statistical models of Thomas–Fermi–Dirac not applicable. However temperatures are generally above the characteristic Debye temperatures of the materials involved so that classical analysis can be used to describe their vibrational properties.

Studies of wave propagation measurements in rocks show that the velocity of compressional waves depends principally upon density and mean atomic weight. Since most common rocks have mean atomic weights close to 21 or 22 regardless of composition, rocks and minerals of very different composition may have the same densities and seismic velocities. Thus it is not easy to infer chemical composition from seismic data alone, and laboratory experiments at the conditions of pressure and temperature that exist deep within the Earth are necessary.

The pioneering experimental work of P. W. Bridgman up to pressures of 100 kbar corresponds to a depth of only 300 km within the Earth. However in the last few years, dynamic determinations of the compressibility of minerals and rocks have been made by a number of workers up to pressures in excess of those at the centre of the Earth. These high pressures are created for very short time intervals behind the front of a strong shock wave set up by an explosive charge, and are an order of magnitude greater than those which can be obtained by static methods. Details of the experimental procedures may be found in Rice *et al.* (1958).

An equation of state obtained from shock wave experiments usually takes the form of a relationship between shock pressure p, shock-induced density ρ, and internal energy E along a curve called the Hugoniot. The Hugoniot curves, upon reduction, yield pressure–density–temperature states for different materials. One form of the equation of state consists of pressure–density isentropes (i.e. constant entropy curves). Upon differentiation the isentropes yield the seismic parameter $\phi = (\partial p/\partial \rho)_s$. Direct comparison is thus possible between values of ϕ obtained from the seismic velocities V_P and V_S (see Eqn. (1.3)) and values of ϕ measured for rocks and minerals in the laboratory under similar conditions of temperature and pressure.

Two assumptions are made in shock wave experiments—that the measured (p, ρ, E) states are in thermodynamic equilibrium and that the compression for a given pressure is the same as that which would be produced by a hydrostatic pressure of the same magnitude. The first condition is satisfied if thermodynamic equilibrium is attained in about 10^{-7} sec or less. This implies a shock front with a thickness of a few tenths of a mm or less.

The thermodynamic states which are produced behind shock waves are determined from the Rankine–Hugoniot equations which express the conservation of mass, momentum, and internal energy or enthalpy across the pressure discontinuity or shock front. Consider a disturbance (corresponding to a shock front) propagating with a velocity u_s into an undisturbed state defined by pressure p_0, density ρ_0 ($= 1/v_0$) and mass velocity zero (see Fig. 5.1). The shock front is assumed to consist of a time-independent pressure profile. If p_1, ρ_1 and u_p are the pressure, density and mass velocity behind the front, the condition that the mass flux in and out of the shock front are equal is

$$\rho_0 u_s = \rho_1 (u_s - u_p). \tag{5.1}$$

The net force on a unit cross-section of the material between $x = A$ and $x = B$ (see Fig. 5.1) is $p_1 - p_0$. This must equal the time rate of change of momentum for this material, viz. $\rho_0 u_s$ through the shock multiplied by the associated velocity change u_p, i.e.

$$p_1 - p_0 = \rho_0 u_s u_p. \tag{5.2}$$

Finally the power input to a unit cross-section of material between A and B, viz. $p_1 u_p$, must equal the time rate of change of energy for the enclosed material, i.e.

$$p_1 u_p = \rho_0 u_s \left(\tfrac{1}{2} u_p^2\right) + \rho_0 u_s (E_1 - E_0) \tag{5.3}$$

G

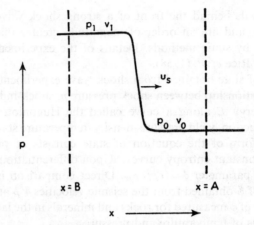

Fig. 5.1. Shock wave pressure profile. The surface $x = B$ moves with the fluid. (After Rice *et al.*, 1958.)

where E_0 and E_1 are the specific internal energies ahead of and behind the shock front respectively. From Eqns. (5.1) and (5.2)

$$u_s = v_0 \sqrt{\frac{p_1 - p_0}{v_0 - v_1}} \qquad (5.4)$$

and

$$u_p = \sqrt{(p_1 - p_0)(v_0 - v_1)}. \qquad (5.5)$$

Hence

$$E_1 - E_0 = \tfrac{1}{2}(p_1 + p_0)(v_0 - v_1). \qquad (5.6)$$

These equations were first derived by Rankine and Hugoniot. Since the specific internal energy of a material is a function of its pressure and density, Eqn. (5.6) may be regarded as the locus of all (p_1, v_1) states attainable by propagating a shock wave into a fixed initial state (p_0, v_0). This locus is defined as the Hugoniot curve centred at (p_0, v_0).

In 1955 Walsh and Christian carried out such calculations for the states produced by 500 kbar shock waves in metals. Since then a large number of papers have been published giving shock wave equation of state data for many metallic elements to pressures, in some cases, up to 9 Mbar (see e.g. Walsh *et al.*, 1957; McQueen and Marsh, 1960; Al'tshuler *et al.*, 1958a, b, 1962). A considerable body of data for compounds as well as for rocks and minerals has been collected in the Compendium of Shock Wave Data (van Thiel *et al.*, 1967).

It must be stressed that in order to interpret the results, the equation of state as determined from shock wave data, which is neither adiabatic nor isothermal, must be reduced to a reference temperature. Temperatures in the shock front are not generally known and additional measurements or assumptions must be made to reduce the pressure–density data to those at absolute zero. Moreover the shock-produced states are characterized by pressure and internal energy—thus an $E(p,T)$ equation of state is required before the data can effectively be used.

It is customary to express the thermal dependence of most equations of state in terms of Grüneisen's parameter γ, where γ represents some average of the volume derivatives of the normal modes. The approximation $\gamma =$ constant is not sufficiently accurate. Knopoff and Shapiro (1969) compared the various conventional methods of computing the temperature and volume dependence of γ and showed that they lead to different isotherms when used in shock wave reduction, and ignore both the contribution of shear modes, which may be ten times as important as the compressional modes, as well as the dispersion of high frequency elastic waves. They further pointed out that lattice models may be of little use in shock wave reduction because the high temperatures associated with the shock probably cause the solid to melt or exceed its elastic limit. Lyttleton (1973) has also pointed out that in the seismic case, harmonic waves of small amplitude are propagated through material already under static pressures in excess of 10^{12} dyn/cm^2 (and at high temperature), whereas in laboratory experiments a shock wave moves into hitherto unstressed material and what is propagated through it is a non-oscillating disturbance in which the peak pressure may attain this same order of magnitude.

5.2. Ramsey's Hypothesis

In 1949 Ramsey proposed that the lower mantle and core have the same chemical composition, the discontinuity at the MCB resulting from a change of mantle silicates to a high-pressure liquid metallic form. This suggestion met with two main difficulties—that of reconciling the large jump in density (by a factor of $\simeq 1\cdot 7$) at the MCB with geochemical theory, and the failure to find positive evidence of such a transformation in shock-wave experiments at the relevant pressures. The materials in the lower mantle are already tightly packed and transformation to a metallic form is unlikely to increase the density. Also, as Birch (1968) pointed out, at one atmosphere the mean atomic volumes of oxides and silicates are less than the mean atomic volumes of the pure metals of which they are made. The transformation to metallic form of mantle material, composed mainly of light elements, must result in a light metal, and metals lighter than chromium ($Z=24$) are all too light for the Earth's core. Figure 5.2 is a plot of the hydrodynamical sound velocity $(\partial p/\partial\rho)_s^{1/2}$ against density along the Hugoniot compression curves for metals up to cobalt and for several rocks. The areas in which the corresponding quantities for the Earth's mantle and core must lie are indicated by the pairs of dotted lines. On the diagram, Ramsey's hypothesis corresponds to a transition from a point such as *A* to a point such as *B*. The experimental evidence clearly shows that this corresponds to a change of atomic number, i.e. of chemical composition, and by a large amount (roughly from 12 to 23). For a given atomic number and given velocity, the rocks are both denser and less compressible than metals. Moreover all experimental data have shown

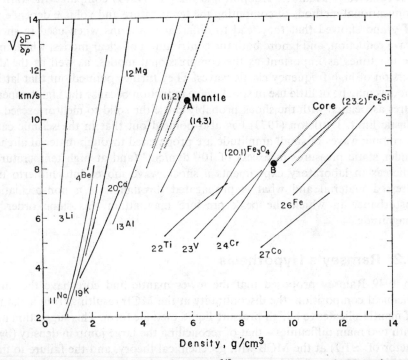

Fig. 5.2. Hydrodynamical sound velocity $(\partial p/\partial \rho)_s^{1/2}$ versus density along the Hugoniot compression curves for metals up to cobalt and several rocks. For a few materials the shock wave velocity has been plotted instead of the adiabatic sound velocity. Atomic numbers or representative atomic numbers (in parentheses) are attached to each curve. The line marked (11.2) shows values for Twin Sisters dunite, a rock composed mainly of olivine (92%) of composition about Fa_{10}; the line marked (14.3) shows values for hortonolite dunite, composed mainly of olivine (90%) of composition about Fa_{50}; the line marked (20.1) is for magnetite; the line marked (23.2) is for an iron–silicon alloy having nearly the composition Fe_2Si. Similar data for a large number of other rocks fall between the lines for the dunites, but have been omitted for the sake of clarity. The areas in which the corresponding quantities for the Earth's mantle and core must lie are indicated by the pairs of dotted lines. Several oxides, silica, periclase and corundum have been compressed to Mbar pressures and also fall in the "mantle" area. (After Birch, 1968.)

that the densities of rocks when compressed well beyond the pressure at the MCB, fall on a smooth continuation of the curves for lower pressures and give no indication of transforming to core densities.

Liu (1974) has constructed a more detailed velocity density plot than Birch's figure (Fig. 5.2) for all solid elements having densities in the range $1 \cdot 5$–8 g/cm^3. Without considering information on the chemical abundances of the elements in the solar system, Liu found that a "Birch" diagram does not unambiguously distinguish the elements of the iron group from other heavy elements as candidates for the core. Elements Ga (31), Ge (32), As (33), Y (39), Zr (40) and Ba (56) are all possible candidates. Birch himself (personal communication) further commented that elements lighter (i.e. of lower atomic weight or number) than V (23) are not likely candidates for the core.

5.3. Bullen's (k, p) Hypothesis

Bullen found for his Earth Model A that there was no noticeable difference in the gradient of the incompressibility, dk/dp, between the base of the mantle and the top of the core. Moreover there was only a 5 per cent difference in the value of k across the MCB. These features are in marked contrast to the large changes in density and rigidity at the boundary. Because of the smallness in the change in k across the MCB and because this change (a diminution) is opposite in sign to that predicted theoretically, Bullen (1949, 1950) suggested another Earth Model B in which he assumed that k and dk/dp are smoothly varying functions throughout the Earth below a depth of about 1000 km. This hypothesis, called the compressibility pressure (k, p) hypothesis, implies that at high pressures the compressibility of a substance is independent of its chemical composition.

On the basis of this hypothesis, Bullen found that there must be a concentration of more dense material near the base of the mantle (region D''). This material could be a mixture of metallic iron with silicates near the MCB or an iron sulfide phase at the base of the mantle. Again if the entire core is liquid so that $V_S = 0$, then from Eqn. (1.3), V_P is given by

$$V_P^2 = k/\rho. \tag{5.7}$$

Jeffreys' original P velocity distribution (see Fig. 1.3) showed a discontinuous jump across the boundary between regions F and G and, to accommodate this, k would have to increase by 32 per cent—excluding the highly improbable case that the density decreases with depth. With Gutenberg's velocity-depth curve (see Fig. 1.3) the effective increase in k would be 23 per cent. On the other hand, as Bullen first pointed out in 1946, if the inner core G is solid and thus capable of transmitting S waves, Eqn. (5.7) is replaced by Eqn. (1.3) in G and the increase in V_P can be accounted for without violating his (k, p) hypothesis. Bullen and Haddon (1967) and Haddon and Bullen (1969) have since revised Bullen's original model B and constructed a series of new Earth models based on the (k, p) hypothesis (see Section 1.4).

The high pressure work of Ahrens *et al.* (1969) on magnesium and aluminum oxides indicates that along an adiabat over the pressure range in the lower mantle $(\partial k/\partial p)_s$ is not constant but decreases markedly. Lattice dynamical calculations also predict that $(\partial k/\partial p)_s$ should decrease as the pressure increases. Moreover such a prediction is given by quantum mechanical calculations of equations of state (such as the Thomas–Fermi–Dirac equation) and general theories of finite deformation (such as the Birch–Murnaghan equation). The results of shock wave experiments and theoretical considerations also indicate quite clearly that k and $(\partial k/\partial p)_s$ do not depend on pressure in the same way for all materials. The behaviour of solids in general is thus inconsistent with Bullen's (k, p) hypothesis, although it is a fairly good approximation in the case of the Earth. This is in part fortuitous—the compressibilities of iron oxide, stishovite, aluminum oxide and magnesium oxide are similar at pressures comparable to those at the MCB. Again the gradients of the bulk modulus of such substances are not constant, although the average values are not so very different from the mean terrestrial value. If the mantle covered a greater range of pressures, the agreement would become less close.

Bullen (1963) also investigated the chemical inhomogeneity of the Earth, where chemical inhomogeneity is used here to include also inhomogeneity arising from phase changes. Assuming hydrostatic pressure (Eqn. (1.6)) and an adiabatic temperature gradient, we have (by definition of ϕ)

$$k = \rho\phi$$

so that

$$\frac{dk}{dp} = \phi\frac{d\rho}{dp} + \rho\frac{d\phi}{dp} = \phi\frac{d\rho}{dp} - \frac{1}{g}\frac{d\phi}{dr}.$$

For a chemically homogeneous region, $\dfrac{d\rho}{dp} = \dfrac{\rho}{k}$ (by definition), and thus

$$\frac{dk}{dp} = 1 - g^{-1}\frac{d\phi}{dr}. \tag{5.8}$$

At any point of a region, whether chemically homogeneous or not, at which $d\rho/dr$ can be assumed to exist, we can write

$$\frac{dk}{dp} = \frac{-\phi}{g\rho}\frac{d\rho}{dr} - \frac{1}{g}\frac{d\phi}{dr}$$

i.e.

$$\frac{d\rho}{dr} = \frac{-\eta g\rho}{\phi} \tag{5.9}$$

where

$$\eta = \frac{dk}{dp} + g^{-1}\frac{d\phi}{dr}.$$ (5.10)

When $\eta = 1$, Eqn. (5.9) reduces to the Adams–Williamson equation (Eqn. 1.8), so that η is a measure of the departure from chemical homogeneity—it is the ratio of the actual density gradient to the gradient that would be obtained if the composition were uniform. An excess temperature gradient normally reduces, while chemical inhomogeneity increases, the value of η.

From Eqn. (5.10) it can be seen that η depends on dk/dp, g and $d\phi/dr$. On Bullen's compressibility–pressure hypothesis, dk/dp is slowly varying and lies between about 3 and 6 throughout most of the Earth's deep interior. Uncertainties in estimates of g are not large, while values of $d\phi/dr$ are immediately obtainable from the P and S wave velocity distributions. It follows from Eqn. (5.10) that η can be estimated in most parts of the Earth's deep interior within limits which can be assigned. Thus it is possible to estimate the degree of departure from chemical homogeneity in any given region and to assess density gradients where the Adams–Williamson equation cannot be used. In a later paper, Bullen (1965a) refined Eqn. (5.10)—in particular he investigated the implications of the variation of k with composition and of the deviation of dk/dp from $(\partial k/\partial p)_{\text{constant composition}}$.

In the lower 200 km of the mantle (region D''), the seismic velocity distributions of both Jeffreys and Gutenberg indicate that $d\phi/dr \simeq 0$, so that $\eta \simeq dk/dp$. Bullen's Model A gives $dk/dp \simeq 3$ in D'' indicating that the lower 100–200 km of the mantle is inhomogeneous. The inhomogeneity is not too great, however —with $\eta = 3$ it contributes only an increase of 0·2 g/cm³ in density across D''. Using more recent values of the seismic velocities (including a decrease in D''), Bolt (1972) obtained a value of 4·7 for η leading to an increase in density of 0·33 g/cm³ through D''.

Table 5.1 gives the density gradients in D'' for a number of recent Earth models. $\eta = 3·1$ for model B1 of Jordan and Anderson (1974) and 2·3 for model UTD 124A′ of Dziewonski and Gilbert (1972). The density gradients in both these models correspond approximately to adiabatic compression ($\eta = 1$). A very slight increase in the gradient of V_P would result in a value of η closer to unity.

Using the data of Taggart and Engdahl (1968) on PcP times, Muirhead and Cleary (1969) devised a variant of the Haddon and Bullen model HB1 with a core radius of 3478 km and with V_S decreasing linearly from the Jeffreys value of 7·25 km/sec at 2700 km to 6·8 km/sec at the MCB. The resultant model, designated ANU1, was found to fit the free oscillation modes reasonably well. Model ANU2 is a variation of model ANU1 in which the thickness of D'' is reduced to 100 km and the values of V_P and V_S at the top of the layer are taken

Table 5.1. Density gradient in D'' for various models, from Bullen's k, p hypothesis. (After Cleary 1974.)

	Radius r (km)	Depth z (km)	V_P (km/sec)	V_S (km/sec)	ρ model (g/cm³)	$d\rho/dz$ (model) (g/cm³/km $\times 10^{-4}$)	$d\phi/dz$ (km/sec² $\times 10^8$)	n^*	$d\rho/dz$** (calc.) (g/cm³/km $\times 10^{-4}$)
UTD124A'									
Top	3611	2760	13·615	7·226	5·382	4·8	0·95	2·3	11·5
Bottom	3482	2889	13·691	7·270	5·444				
B1									
Top	3625	2746	13·63	7·22	5·50	5·2	0·14	3·1	15·5
	3510	2861	13·67	7·27	5·56	8·0	0·0	3·2	16·0
Bottom	3485	2886	13·67	7·27	5·58				
Bolt									
Top	3621	2750	13·63	7·30	(5·6)	(20·6)	−1·52	4·7	23·5
Bottom	3475	2896	13·33	6·99	(5·9)				
ANU2									
Top	3582	2789	13·68	7·24	(5·5)	(20·0)	−1·32	4·5	22·5
Bottom	3482	2889	13·34	6·82	(5·7)				
B2									
Top	3550	2821	13·66	7·33	5·52	80·0	13·88	−10·0	−50·0
Bottom	3485	2886	13·46	6·55	6·04				

* Based on $\eta = dk/dp - g^{-1}\, d\phi/dz$, where $dk/dp = 3\cdot2$.

** Based on $d\rho/dz = \eta g\rho/\phi$, where $g\rho/\phi \approx 5 \times 10^{-4}$ g/cm³/km in D''.

from the models of Hales and Herrin (1972) and Hales and Roberts (1970) respectively while the values at the bottom are those given by "diffracted" wave studies for a core radius of 3482 km. The calculation gives $\eta = 4\cdot5$. Jordan's (1972) model B2 has an unjustifiably low value of V_S at the MCB leading to a change of sign in the density gradient between that of the model and that calculated from Eqn. (5.9) and a negative value $(-10\cdot0)$ for η.

In the transition region F between the outer and inner core, Jeffreys' velocity distribution (characterized by a large negative P velocity gradient, i.e. $d\phi/dr \gg 0$) leads to a value of η of 38 entailing a density increase of the

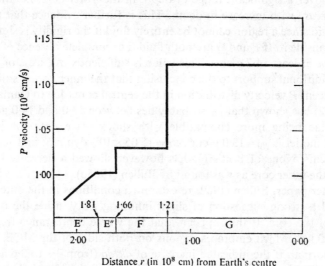

Distance r (in 10^8 cm) from Earth's centre

Fig. 5.3. P velocities and the regions E', E'', F and G of the lower core. (After Bullen, 1965b.)

order of 3 g/cm³ through F. On the other hand, Gutenberg's velocity distribution gives large negative values of $d\phi/dr$ so that η is significantly less than unity (actually negative) implying an unstable distribution of mass. It would seem that seismic velocity gradients much in excess of those in regions D and E cannot exist in the Earth's deep interior. An infinite gradient (i.e. a velocity discontinuity) on the other hand is not impossible since then the range of depth of any instability would be zero.

In Bolt's (1962, 1964) revision of Jeffreys' distribution of the velocity of P waves in the deep interior of the Earth, the core is divided into four regions E', E'', F, and G.* The velocity distribution down to the bottom of E' is the same as that of Jeffreys for corresponding depths inside E. There are discontinuous jumps in V_P at the E''–F and F–G boundaries and V_P is constant

* Bolt's reinterpretation has since been superseded (see Sec. 1.2).

in E'', F and G (see Fig. 5.3). Birch (1963), using shock wave data at pressures of the order of 10^6 bar, inferred that the density ρ_0 at the centre of the Earth does not exceed 13 g/cm³. Bullen (1965b) investigated the consequences of this limiting value of ρ_0 using Bolt's distribution of V_P in the core (Fig. 5.3). Since ρ is not likely to decrease with depth in the core, Bullen's (k, p) hypothesis implies, through Eqn. (1.1) that departures from smooth variations of V_P with r are accompanied by similar departures in the variation of μ rather than of k. Bullen showed that it is impossible for ρ_0 to be as low as 13 unless there is substantial rigidity in both regions F and G. In addition Bullen found that $d\mu/dr > 0$ over a significant range of depth in the lower core, i.e. the rigidity must *decrease* with increase in depth. This is further evidence that the inner core is solid since a region cannot be entirely fluid if the rigidity is significantly changing inside it. If F and G are both fluid (i.e. complete absence of rigidity), ρ_0 must be at least 14·7 g/cm³. This value is sufficiently in excess of 13 g/cm³ to give additional support to the conclusion that the inner core is solid. Using a more recent P velocity distribution in the central core (due to Qamar, 1971), Bolt (1972) has shown that ρ_0 probably lies between 13·0 and 14·0 g/cm³, the lower value being more compatible with shock wave data for iron. His preferred model is $\rho_0 = 13\cdot0$ g/cm³, $k_0 = 15\cdot05 \times 10^{12}$ dyn/cm² and $\mu_0 = 1\cdot25 \times 10^{12}$ dyn/cm². None of Bolt's models however showed a decrease in μ with depth in the inner core as was found by Bullen (1965b).

In a later paper, Bullen (1969) re-examined conditions in the outer core E. He found a strong suggestion of slight inhomogeneity inside the outermost 700 km of the core. It thus appears that the whole depth range from about 2700–3600 km, which embraces regions on both sides of the MCB, is some-what abnormal. If the indicated increase of dk/dp (from 3·1 to 3·6 across the outermost 700 km of the core) is due solely to compositional changes, such changes might be enough to inhibit convection currents and confine convection in the core to the depth range $3600 \sim 4500$ km. A range of depth of the order of only 1000 km might well be insufficient to satisfy the requirements of the dynamo theory of the Earth's magnetic field (Section 4.2). Bullen thus tentatively suggested that there may be a continuous phase change near the top of the core, the rate of change diminishing with depth—the lower part of the outer core (below a depth of about 3600 km) being probably nearly uniform in composition and phase ($\eta \simeq 1$ and $dk/dp \simeq 3\cdot6$).

Using values of $d\phi/dz$ computed for his model KOR5 (see Fig. 1.6), Qamar (1973) found, from Eqn. (5.10), that $\eta \simeq 1$ in region E. In F, the small velocity gradient leads to a value of η in the range $3 \sim 4$ indicating a possible change in phase or chemical composition. The large velocity gradient at the top of region G in model KOR5 has interesting consequences. Assuming that dk/dp has about the same value as in the outer core, the large value of dV_P/dr leads to negative values of η if $V_S = 0$. With $\eta = 1$, $dk/dp = 4$ in G (Bullen, 1963) and $V_S \simeq 3.5$ km/sec near the Earth's centre (Dziewonski and Gilbert, 1972), Qamar found that V_S must increase rapidly with depth by about 0.3 km/sec from the inner core boundary to about $r = 950$ km and then slowly decrease by about 0.15 km/sec towards the Earth's centre.

5.4. Bullen's Fe₂O Hypothesis

Bullen (1973a, b) has developed an alternative model of the cores of the terrestrial planets which avoids the main difficulties of the phase-transition theory whilst retaining the important feature that the pressure p_c at the MCB is critically involved in the change of properties there. His theory is based on a suggestion by Soroktin that the Earth's outer core consists of Fe₂O. Soroktin calculated that this oxide, which is unstable at ordinary pressures, becomes stable at the pressures in the Earth's core and has a density–pressure relationship in agreement with that in the Earth's outer core. However, whereas Soroktin attributes the occurrence of Fe₂O in the outer core to a breakdown of FeO into Fe₂O and oxygen, Bullen associates it with the equation Fe₂O ⇌ FeO + Fe. Bullen considers a model family of planets with the following properties. All planets of the family are composed of two primary materials—a basic mantle material X, and Fe₂O. (The composition of X is not specified, but it is likely to contain some FeO.) In all the planets, the ratio of the mass of X to the mass M of the planet is the same. In those planets which contain Fe₂O, the Fe₂O occurs as a distinct zone (the outer core), throughout which $p \geqslant p_c$. In those planets where $p < p_c$ in an Fe₂O zone, some or all of the Fe₂O has broken down into FeO and Fe. This FeO (which Bullen calls Y) forms part of the mantle and is additional to any FeO that may be part of X. The Fe falls to form an inner core.

Bullen's family of models has three sub-sets which he calls H, J, K (see Fig.

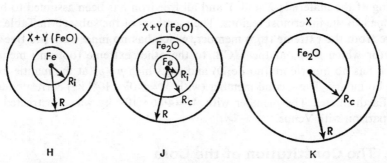

Fig. 5.4. Materials in the interiors of the three sub-sets (H, J, K) of the terrestrial planets. The outermost zones, all containing the material X, are mantles. The Fe₂O zones are referred to as "outer cores", the Fe zones as "inner cores". (After Bullen, 1973a.)

Table 5.2. Structures of model planets of the J subset (possessing both outer and inner cores). (After Bullen 1973b.)

Mass (M) (10^{24} kg)	Radius (R) (10^3 km)	Mean density ($\bar{\rho}$) (g cm^{-3})	Mantle thickness (10^3 km)	Mantle density (g cm^{-3})	Outer-core thickness (10^3 km)	Inner-core radius (10^3 km)
6·29	6·44	5·62	2·83	4·47	3·61*	0
6·03	6·37	5·57	2·89	4·50	2·28	1·20
5·96	6·35	5·57	2·91	4·51	2·13	1·31
5·65	6·26	5·50	3·00	4·55	1·62	1·64
5·34	6·16	5·45	3·09	4·60	1·21	1·86
5·04	6·07	5·38	3·21	4·65	0·83	2·03
4·86	6·01	5·34	3·29	4·68	0·60	2·12
4·75	5·97	5·32	3·34	4·69	0·46	2·17
4·46	5·87	5·26	3·52	4·73	0·07	2·28
4·36	5·83	5·22	3·54	4·74	0	2·29

* Radius.

5.4). The subset H includes the smallest planets which have no Fe_2O zones and thus no outer cores: they have mantles composed of X and Y, and inner cores of Fe. The sub-set K includes the largest planets which have mantles composed purely of X, outer cores of Fe_2O, and no inner cores. The subset J consists of intermediate planets which have mantles composed of X and some Y, outer cores of Fe_2O, and inner cores of Fe. Earth and Venus correspond to members of J, Mars of H. No known planets correspond to K.

Table 5.2 gives the results which Bullen has computed for a number of members of the subset J. For a first approximation, compressibility effects have been neglected (and variations of density inside mantles, outer and inner cores), as well as possible volume changes and/or chemical interactions in the mixing of the materials X and Y and all free iron has been assumed to have dropped to the innermost regions. The members of the subset J in Table 5.2 range from the extreme (top) member, which has no inner core (no free Fe) and for which $p = p_c$ at the MCB, to the other extreme (bottom) member which has no outer core (no Fe_2O) and for which $p = p_c$ at the mantle-inner core boundary. The second member (with $M = 6·03 \times 10^{24}$ kg) corresponds to the Earth model. The member with $M = 4·86 \times 10^{24}$ kg was constructed for comparison with Venus.

5.5. The Constitution of the Core

As already pointed out an iron–nickel core has too large a density and too small a bulk sound speed to be compatible with geophysical data—shock wave experiments by McQueen and Marsh (1966) on iron–nickel alloys indicate that pure iron is about 8 per cent denser than the outer core. A lighter alloying element that would increase the bulk sound speed is required. Limitations upon possible choices are that the element be reasonably abundant, miscible with liquid iron, and possess chemical properties which would allow it to enter the core. Possible candidates are H, He, C, O, N, Mg, Si and S. Ringwood (1966a) has pointed out that H, He, C, O and N may be rejected since they are known to form interstitial solid solutions with iron. Additions of these elements would not significantly decrease the density of iron, since they occupy holes already present in the lattice. Magnesium is unlikely to be present in any amount since it has a much greater affinity for oxygen than has silicon, i.e. any chemical conditions that may have led to the incorporation of magnesium in the core would have caused the incorporation of much larger amounts of silicon. This leaves just silicon and sulphur as possibilities. The case for sulphur will be discussed later—Ringwood has given strong arguments (1966b) against there being any appreciable amounts of sulphur in the core. By a process of elimination he thus favours silicon as the most likely extra component of the Earth's core. He also pointed out that the presence of substantial quantities of silicon in the metal phase of enstatite chondrites shows that chemical conditions during the formation of the solar system were, at least in some regions, favourable for the reduction of silicates. Ringwood also showed that an Earth model constructed from the abundances of non-volatile elements in Type I carbonaceous chondrites also requires the presence of silicon in the core. The chemical composition of such an Earth model is given in Table 5.3.

In Ringwood's model the terrestrial planets formed in an initially cold gas and dust cloud. The Earth is believed to have accreted from primitive oxidized dust similar in composition to the material of Type 1 carbonaceous chondrites. In this model, reduction of the iron oxide by organic matter in the dust begins to occur inside the growing Earth due to the heat liberated by the release of gravitational potential energy. As the temperature of the Earth increases, the metallic iron thus produced melts and sinks to the centre of the Earth to form the core. As a consequence of this high temperature reduction process, some Si enters the metal phase, accounting for the low density of the core, and some S is lost from the Earth by degassing along with other volatiles which are believed to be depleted in the Earth relative to Type 1 carbonaceous chondrites. In the final stages of the accretion of the Earth, a hot, dense silicate atmosphere builds up which later escapes from the Earth and condenses into a sediment-ring from which, Ringwood believes, the moon accreted.

Ringwood's model has encountered a number of difficulties. If the core of the Earth is formed *in situ* by reduction, the reaction products, H_2 and CO,

Table 5.3. Composition of Earth as Derived by Reduction from
Composition of Type I Carbonaceous Chondrites.
(After Ringwood 1966b.)

	1	2	3	4
SiO_2	33·32	35·85	29·84	43·25
MgO	23·50	25·19	26·29	38·10
FeO	35·47	6·14	6·38	9·25
Al_2O_3	2·41	2·59	2·69	3·90
CaO	2·30	2·47	2·57	3·72
Na_2O	1·10	1·18	1·23	1·78
NiO	1·90	—	—	—
	100·00	73·52	69·00	100·00
Fe		24·88	25·87	
Ni		1·60	1·66	
Si		—	3·47	
		26·48	31·00	

Column 1. Average composition of principal components of Type I carbonaceous chondrites (Orgueil and Ivuna) on a C-, S- and H_2O-free basis (analyses by Wiik, 1956).

Column 2. Analysis from Column 1 with (FeO/(FeO + MgO)) reduced to be consistent with probable value for earth's mantle (0·12).

Column 3. Analysis from Column 2 with sufficient SiO_2 reduced to elemental silicon to yield a total silicate to metal ratio 69/31 as in the earth.

Column 4. Model mantle composition: silicate phase from Column 3 recalculated to 100 per cent.

will form an enormous atmosphere totalling perhaps half the mass of the core. No trace of such an atmosphere remains and an efficient dissipation mechanism must be postulated. Again the H_2O/H_2 and CO_2/CO ratios in the gases which degas from the mantle far exceed those which would be in equilibrium with metallic iron and ferromagnesian silicates.

The presence of silicon in the core whereas the mantle contains substantial quantities of oxidized iron implies that the mantle is not in chemical equilibrium with the core. Ringwood regards this as an important constraint on any process of core formation. In particular it is incompatible with the theory that the material from which the Earth accreted was composed of an intimate mixture of silicate and metal particles similar to ordinary chondrites. The disequilibrium between mantle and core has been discussed in more detail in earlier papers by Ringwood (1959, 1961). The deep interior of the Earth is initially highly oxidized and rich in volatiles, whereas the outer regions are progressively more reduced and poor in volatile components. After melting near the surface, the metal phase, consisting of an iron-nickel-silicon alloy collects into bodies which are large enough to sink into the core.

In Ringwood's model, core-mantle separation took place so rapidly that chemical equilibrium was not attained. In addition, the Fe^{+2}/Fe^{+3} ratio of the mantle is a factor of 30–40 below that which would be in equilibrium with metallic iron at 2500°C, the mean temperature believed by Ringwood of the interior of the Earth during the metal-silicate separation. Again the abundance of Ni in the upper mantle, as measured in ultramafic rocks, is nearly two orders of magnitude higher than would be expected if equilibrium partitioning of Ni between metal and silicate phases had occurred during core formation. Ringwood thus believes that the mean oxidation state of the mantle is such that it is not now, nor has it ever been, in equilibrium with metallic iron. Clark, Jr. *et al.* (1972) have shown, however, that the diffusion coefficients for Ni in olivine (Clark and Long, 1971) imply that complete removal of Ni from 2 mm olivine grains is possible in about ten years at the zero pressure melting-point of iron, 1805°K. This, coupled with the high Ni content of mantle rocks, places severe limits on the time scale of the coalescence of small metallic droplets into the larger pools which sink to the centre of the Earth. Equilibrium between metal and silicate can only be obtained by diffusion across the interface of the sinking bodies of metal. If the rate of sinking of metal is high compared to the rate of attainment of equilibrium by diffusion (which Ringwood believes to be the case), the core which separates will not be in equilibrium with the mantle. After the core has separated and the mantle is molten, both regions will become homogenized by convection, but they will remain out of equilibrium with one another.

Some of the difficulties of Ringwood's model have been removed by Murthy and Hall (1970) who proposed a homogeneous accretion model in which the composition of the Earth is 40 per cent carbonaceous chondrite, 50 per cent ordinary chondrite and 10 per cent iron meteorite. Since all the iron which eventually enters the core already exists in the reduced state in this model, no enormous atmosphere of reaction products is built up and there is, therefore, no difficulty encountered in its dissipation. Again since this mixture of meteorite types contains abundant Fe and FeS grains, melting first begins at a temperature close to the Fe–FeS eutectic, 550° lower than the melting-point of pure Fe at zero pressure. Because the Fe–FeS eutectic temperature is relatively insensitive to pressure (Brett and Bell, 1969), temperatures in the Earth keep the resulting Fe–S melt liquid until it reaches a depth well within the present inner core–outer core boundary, thus avoiding the refreezing difficulty in Ringwood's model.

The recent discovery of Xe^{129} produced by the decay of the short-lived extinct radioactive I^{129} ($t_{\frac{1}{2}} = 17 \times 10^6$ yr) by Buolos and Manuel (1971) in deep mantle gases sets limits, not only on the time of formation of the Earth relative to nucleosynthesis, but also on the degassing history of the outer parts of the Earth. Specifically, this precludes a sustained high temperature regime for the mantle after a few mean life times (≈ 80 m yr) of this extinct radio-nuclide. If the Earth accreted rapidly ($10^3 - 10^4$ yr), it is doubtful whether it would cool to low enough temperatures to retain Xe^{129} in this short a time. Hot accretion models for the Earth (Ringwood, 1966a, b; Anderson and Hanks, 1972) in which an efficient trapping of the gravitational potential energy occurs, seem untenable in the light of this discovery (but see also Section 2.4).

Traces of the fission-produced Xe isotopes 131, 132, 134, and 136 were also found by Buolos and Manuel. The relative isotopic compositions of these isotopes strongly suggest that a mixture of spontaneous fission by U^{238} and Pu^{244} has been responsible for their production. These observations give direct evidence that the formation of the Earth did not appreciably postdate the formation of the meteorites. If it did, isotopic anomalies from the previous decay of extinct radioactivities should almost certainly have been obliterated well in the formation of the planet.

Murthy and Hall (1972) thus maintain that models for the early history of the Earth must simultaneously be in accord with an essentially low temperature accretion and yet lead to the formation of the core nearly simultaneously with, or very soon after, accretion. In any evolutionary model of core formation, this restriction effectively rules out pure Fe–Ni melting and sinking into the interior. Murthy and Hall therefore investigated low melting components (in particular sulphur) that would satisfy the geophysical data of the core as well as a number of geochemical considerations of the mantle and core.

The lowest melting composition for any initial Earth is probably in the binary eutectic system Fe–FeS. The eutectic melting temperature of this system at zero pressure is 988°C. Additional components such as FeO, C, Ni would amost certainly reduce this value. Brett and Bell (1969) have shown that the Fe–FeS eutectic is little affected by pressure. This has been confirmed by Ryzhenko and Kennedy (1973) who carried out experimental work on the system Fe–FeS up to pressures of 60 kbar. They found that the temperature of the eutectic rises to a broad maximum of 1015°C in the 30–40 kbar region and then decreases to $\sim 1005°$ at 60 kbar. The composition of the eutectic shifts with pressure in the iron-rich direction. If this trend continues, the melting temperatures of the eutectic must begin to rise again to approach that of pure iron. The curve that best fits the experimental data of Ryzhenko and

Kennedy which shows a broad maximum in the 30–40 kbar region, should also show a broad minimum at some pressure above 60 kbar before it begins to rise again. The authors offer no explanation for this strange behaviour.

In general, the effect of pressure on the eutectic temperature is less than on the melting point of the end members (Newton *et al.*, 1962) because of the large entropy of mixing. Thus it can be expected that in the interior of the Earth, the difference between the melting point of iron and an Fe–FeS melt would be even larger than that at zero pressure. At 1 atm, the difference in melting temperatures of pure Fe and the eutectic Fe–FeS is 550°C. From a consideration of the behaviour of this eutectic at higher pressures, Hall and Murthy have shown that this difference may be as much as three times higher in the core.

The initial temperatures at zero time for a number of time scales of accretion are shown in Fig. 5.5 (after Murthy and Hall, 1972). It can be seen that in all the models, melting temperatures of the Fe–FeS eutectic are exceeded for the bulk of the Earth's mass but are well below the melting point of Fe. If core formation practically simultaneous with accretion is a significant

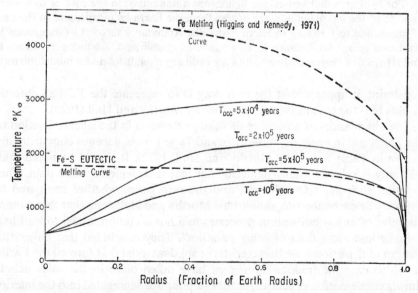

Fig. 5.5. Temperature profiles for the Earth for time scales of accretion ranging from 5×10^4 to 10^6 yr, schematically shown from Hanks and Anderson (1969). The lower three profiles are for rapid accretion of material maintaining low surface temperatures (300°K). The uppermost profile assumes an initial high nebular condensation temperature and radiative equilibrium during accretion and adiabatic compression in the interior for a short time scale (5×10^4 yr) accretion. (After Murthy and Hall, 1972.)

Table 5.4. Origin of the core. Volatile element abundance patterns in Earth and meteorites. (After Murthy and Hall 1972.)

Element	Abundance in crust and mantle*)	Abundance factor I	Abundance factor II	Abundance factor III
S	605	$1\cdot2\times10^{-3}$	$6\cdot0\times10^{-3}$	$2\cdot4\times10^{-3}$
C	1590	$2\cdot0\times10^{-3}$	$0\cdot15$	$0\cdot5\times10^{-2}$
N	230	$4\cdot7\times10^{-3}$	$0\cdot29$	$1\cdot0\times10^{-2}$
H (as H_2O)	$1\cdot2\times10^5$	$2\cdot2\times10^{-2}$	$1\cdot3$	$6\cdot0\times10^{-2}$
F	1240	$0\cdot34$	$1\cdot13$	$0\cdot62$
Cl	250	$0\cdot13$	$6\cdot5$	$0\cdot31$
Br	4	$0\cdot19$	$9\cdot5$	$0\cdot45$
I	$0\cdot31$	$0\cdot22$	$5\cdot5$	$0\cdot52$
^{20}Ne	$2\cdot7\times10^{-4}$	$0\cdot12$	$4\cdot3$	$0\cdot31$
^{36}Ar	$5\cdot5\times10^{-4}$	$0\cdot09$	$2\cdot3$	$0\cdot21$
^{84}Kr	$1\cdot1\times10^{-5}$	$0\cdot10$	$4\cdot6$	$0\cdot25$
^{132}Xe	$3\cdot9\times10^{-7}$	$3\cdot2\times10^{-3}$	$0\cdot15$	$8\cdot1\times10^{-3}$

* Data from Murthy and Hall (1970). All abundances are in atoms per 10^6 atoms Si. For hydrogen and xenon, the figures are a minimum; in the case of hydrogen, because of the unknown amount of loss from the Earth by escape, and in the case of xenon due to trapping in Earth materials. Abundance factor I = (abundance in crust and mantle/abundance in carbonaceous chondrites). Abundance factors II and III refer to corresponding values for ordinary chondrites and a model mixture.

constraint, it appears that the only way is to segregate the Fe–FeS eutectic liquids into the core as discussed earlier by Murthy and Hall (1970).

The abundances of a number of volatile elements in the silicate fraction of the Earth and in ordinary chondrites and Type 1 carbonaceous chondrites are shown in Table 5.4 (after Murthy and Hall, 1972). It should be noted that, not only is sulphur highly depleted, but that it is depleted more than other volatiles such as H_2O halogens, and the rare gases, whether compared to ordinary or carbonaceous chondrites. Murthy and Hall state that "we cannot conceive of any volatilization processes in a hot accretion model to lead to a sulphur loss more than of other volatiles". They concluded that either this pattern of depletion was characteristic of solid material that formed the Earth, in which case the fractionation must have taken place in the solar nebula during condensation processes, or that sulphur was segregated into the interior by Fe–FeS melting in the early Earth. The first possibility—cosmochemical fractionation during the condensation of matter in the solar nebula—is unlikely, and Murthy and Hall are forced to accept the second. It must be pointed out, however, that Ringwood (1966a, b) is of the opinion that most of the cosmic abundance of sulphur was not retained by the Earth and that the core contains only a small fraction of the primordial sulphur, which must have been lost from the Earth before or during accretion. This question will be discussed again in Section 5.6.

As soon as the temperature in the accreting Earth exceeds 990°C, the Fe–FeS liquid will form and, because of its higher density, will sink down to form the core. Once the process has started the gravitational energy liberated will be available practically simultaneously with the accretion process, thus satisfying the constraint that the Earth's core be formed at the time of, or very soon after, accretion. The energetics of core formation by iron sinking to the centre has been discussed in Section 2.3 and 2.4. Birch (1965) has estimated that the energy available for heating is about 400 cal/g*. The density of Fe is 7·9 g/cm³ and of the Fe–FeS eutectic about 5·0 g/cm³. Keeping other factors the same, the thermal energy released in Murthy and Hall's model would be approximately 250 cal/g and would result in a temperature increase of ≈ 1000°C. This temperature increase is probably an upper limit because segregation of the core during accretion will entail radiative losses. It can be seen from Fig. 5.5 that even with this additional heat, metallic iron is not likely to melt in a substantial fraction of the Earth and large scale silicate melting and degassing will certainly not occur. This is in keeping with the retention of radiogenic Xe^{129} formed in the mantle during the first few m yr of the Earth's lifetime.

Murthy and Hall thus interpret the mode of formation of the solid inner core as due to segregation of iron from a Fe–Ni–S melt. Since the liquid in equilibrium with a solid in a multi-component system cannot have the same composition as the solid, there must be a compositional difference between the inner and outer core. The presently accepted density of the inner core of about 13·5 g/cm³ (Bolt, 1972) is compatible with an Fe–Ni alloy. A consequence of this mode of formation of the inner core is that the boundary between the inner and outer core be on the liquidus surface of the Fe–Ni–S ternary system. The temperature at the inner core—outer core boundary should thus be less than the melting point of iron, estimated by Higgins and Kennedy (1971) to be 4250°C.

Not only sulphur but substantial amounts of carbon will also be incorporated into the core. The low density of the core is then due to the presence of about 15 per cent of these elements in an Fe–Ni core. A composition of the Earth based on a core containing about 15 per cent S and C and a "pyrolite" (Green and Ringwood, 1963) type of mantle would not correspond to any single class of known meteorites. Appropriate initial compositions can only be represented by a mixture of materials condensed over a range of temperatures in the solar nebula. A satisfactory mixture for the Earth's composition with

---

* In a later paper, Flaser and Birch (1973) revised this value to 590 cal/g.

respect to major oxides, metal phases and volatiles can be obtained from about 40 per cent carbonaceous chondrites, 45 per cent ordinary chondrites and 15 per cent iron meteorites (Murthy and Hall, 1970). These authors do not suggest that these proportions of meteorites actually accreted to form the Earth. The mixture is simply an approximate representation of the bulk Earth material of solar nebular condensates in terms of the major classes of meteorites.

Usselman (1975a) has investigated the liquidus relations of the Fe-rich portions of the Fe–Ni–S system at pressures from 30 to 100 kbar. He found that up to 6·5 wt per cent Ni had very little effect on the melting relations of the Fe–FeS system at and below 80 kbar. He thus extrapolated the Fe–FeS eutectic to higher pressures in order to estimate eutectic temperatures at core pressures, using the compressibility of Fe and the calculated compressibility of FeS in the Kraut-Kennedy (1966) melting relationship. The eutectic temperatures and compositions are about 1800°C and 17·5 wt per cent S at the MCB, and about 2100°C and 15 wt per cent S at the inner core—outer core boundary. In a later paper, Usselman (1975b) used these results to obtain a model of the formation, composition and temperature of the core.

As the core-forming liquid coalesced, it would become gravitationally unstable and sink while the silicate material would be displaced. This process would lead to an initially homogeneous core of Fe–FeS eutectic composition. The high-pressure Fe–FeS eutectic conditions closely approximate the first melting liquid in the Fe–Ni–S system with up to 6·5 wt per cent Ni (Usselman, 1975a). Because of the loss of gravitational potential energy resulting from core formation there would be an increase of temperature. This increase in temperature would cause the incorporation of much of the available metallic Fe into the Fe–FeS liquid, resulting in a liquid richer in Fe than the eutectic composition at the ambient pressure. As the initial thermal profile approaches the present thermal profile, Fe will crystallize out along the Fe-eutectic liquidus and sink gravitationally to begin to form the solid inner core. As the Earth cools, additional Fe will crystallize out of the liquid and enlarge the solid inner core.

The present composition of the outer core may be approximated by the intersection of the temperature gradient with the liquidus surface. Usselman (1975b) has pointed out that the inner core–outer core boundary is probably a chemical boundary as well as a solid–liquid boundary. This is supported by the conclusions of Derr (1969) and Press (1968) that the inner core is essentially pure Fe or Fe–Ni. Bolt (1972) has estimated that there is a density increase of about 1·8 g/cm^3 at the inner core–outer core boundary—such an increase is greater than would be associated with a solid–liquid transition of a homogeneous material.

For the Fe–Ni–S core models (Murthy and Hall, 1970; Anderson *et al.*,

1971) the minimum temperature of the outer core is essentially the Fe–FeS eutectic. If temperatures are extrapolated to the pressure of the inner core–outer core boundary, a temperature of approximately 2200°C is obtained (Usselman, 1975a). Using Higgins and Kennedy's (1971) extrapolation of the melting temperature of pure iron (about 4250°C at the inner core–outer core boundary), the possible range of temperatures at the inner core–outer core boundary lies between 2200°C and 4250°C.

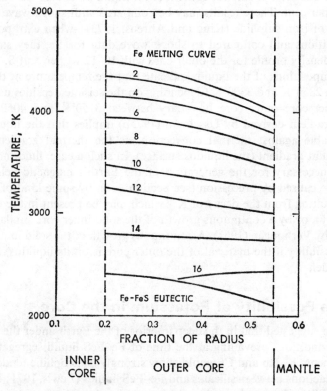

Fig. 5.6. Schematic sketch of the sulphur content of the Fe-eutectic liquidus surface (contoured at constant sulphur content). The sulphur content of the core is determined by the intersection of the contours with the thermal gradient. (After Usselman, 1975b.)

For the Fe–FeS core model, Fig. 5.6 shows the sulphur content of the liquidus surface between Fe and the Fe–FeS eutectic, contoured with depth and temperature. Usselman (1975b) has estimated the densities of various Fe–FeS compositions on the liquidus surface at the MCB and at the inner core–outer core boundary. It is then possible to select the liquidus compositions at these two boundaries which agree with the seismically determined densities of the liquid outer core. If the seismic densities of Haddon and Bullen (1969) are used, the composition of the outer core at the MCB is about 10 wt per cent S, and at the inner core–outer core boundary about 7 wt per cent S. The average sulphur content of the liquid outer core, is about 9 wt per cent. If the densities are chosen from the precision bands given by Bolt (1972) the composition of the liquid outer core at the MCB lies between 13 and 9 wt per cent S, while at the inner core–outer core boundary it lies between 8 and 5 wt per cent S. The average sulphur content of the liquid outer core lies between 11 and 8 wt per cent. These results may be compared with shock wave data (to 400 kbar) of iron sulphide (King and Ahrens, 1973)—when extrapolated to core conditions and compared to shock wave data for Fe, they satisfy the pressure–density profile for the outer core with 10–12 wt per cent S.

The compositions of the liquids indicate that the temperature at the MCB is between $2850°K$ and $3300°K$ (depending on the seismic densities used), and at the inner core–outer core boundary between $3750°K$ and $4050°K$. The thermal gradient derived by Usselman (1975b) implies that the liquid outer core is stable against thermal convection, as the thermal gradient is the melting-point gradient (the liquidus surface). In such a case, fluid motions in the core, necessary for the generation of the Earth's magnetic field, could possibly be caused by precession (see Section 4.7), by some kind of thermal engine resulting from the decay of K^{40} which may be present in the core (see Section 5.6), or by a continuous growth of the solid inner core similar to that proposed by Verhoogen (1961). A continuous growth of the solid inner core is only a possibility if the material of the outer core is on its liquidus, as in the above model.

5.6. The Possibility of Potassium in the Core

It has been suggested that in the deep interior of the Earth under the strongly reducing conditions prevailing at the time of Fe–FeS liquid segregation, the alkali elements K, Rb and Cs would show strongly chalcophilic tendencies in possible reactions between silicates and Fe–FeS liquids (Lewis, 1971; Hall and Murthy, 1971). Thermodynamic calculations by these authors have shown that in a mixture of silicates and Fe–FeS, equilibrations will result in an incorporation of K into the sulphide melt in reasonable concentrations, at temperatures where the silicates remain solid but Fe–FeS exists as a liquid. This suggests that during core formation significant amounts of potassium can be withdrawn into the core. The chalcophilic behaviour of K under conditions appropriate to the primitive Earth has been verified by recent laboratory experiments (Goettel, 1972). Moreover the incorporation of K into the core can proceed under less strongly reducing conditions than those required to introduce elemental silicon into the core.

Hall and Murthy (1971) showed that during core formation, it is possible to separate K from U and Th. Differential segregation of K into the core, leaving behind the initial complement of U and Th in the mantle and the crust would also be consistent with the low K/U ratio of terrestrial rocks ($\sim 1 \times 10^4$) relative to chondritic meteorites ($\sim 7 \times 10^4$) (see Wasserburg *et al.*, 1964).

The presence of K^{40} in the core has many implications for the thermal history of the core as well as the lower mantle. For the mixture chosen by Murthy and Hall (1972), the total K content would be about 500 ppm. Assuming that about three-quarters of this has been segregated into the core, the K content, about 375 ppm, would lead to a present-day heat generation in the core of about 2×10^{19} erg/sec. Thus a significant amount of thermal energy due to radioactivity should exist in the core. If the solid inner core was formed by the segregation of pure metal from an Fe–FeS melt, and is in chemical equilibrium with the outer liquid core, the outer core will be the site of the K^{40}. The heat produced by K^{40} in the outer liquid core would thus be available to set up convective motions. Estimates of the total ohmic dissipation by currents in the core to maintain the geomagnetic field are of the order of 5×10^{16} erg/sec. Thus the present estimate of heat generation due to K^{40} in the core (2×10^{19} erg/sec) is more than sufficient for the generation of the geomagnetic field even at very low thermodynamic efficiencies ($\approx 0 \cdot 01$).

The hypothesis that a large fraction of the Earth's K is in an Fe–FeS core has been disputed by Oversby and Ringwood (1972). They first present data on the distribution of K in meteorites. The distribution of K in the Abee enstatite and in two ordinary chondrites has been investigated by Shima and Honda (1967). On a strict analogy between Abee and the Earth, a maximum of 2·5 per cent of the Earth's K could be in the core—the ordinary chondrites analysed by Shima and Honda had less than 2 per cent K in non-silicate phases. Oversby and Ringwood also measured the distribution of K between a synthetic basalt and an Fe, FeS metallic phase containing 28 per cent S. In their first run they found a minimum distribution coefficient for K between silicate and metal of 25, limiting the amount of K in the Earth's core to less than 2 per cent of the total amount available in the whole Earth. In their second run, they obtained a minimum distribution coefficient of 50, corresponding to a maximum of 1 per cent of K in the Earth being in the core. The minimum distribution coefficients result from their inability to detect any K in the metal phase. Moreover their experiments contained a much higher FeS/Fe ratio than would be possible for the Earth—this should have favoured entry of K into the metal phase. Their work points out the dangers of making predictions based upon thermodynamic calculations for which pure oxides are used as analogs of the actual silicate phases involved in the reaction of interest.

The paper by Oversby and Ringwood (1972) brought forth objections from Goettel and Lewis (1973) to which Oversby and Ringwood (1973) replied.

No attempt will be made to produce the arguments and counter-arguments to the claim that a large proportion of the Earth's potassium is in the core. One point that Goettel and. Lewis make is that the chemical and physical conditions which are relevant to the partitioning of K between the Fe–FeS core and the silicate mantle and crust are the conditions which existed during the primary differentiation of the chondritic Earth into core and mantle and not the conditions in the present crust and mantle. They therefore argue that the partitioning experiments of Oversby and Ringwood using highly differentiated basaltic material are not necessarily relevant to the distribution of K in a primitive, differentiating, chondritic Earth. Oversby and Ringwood mention further experiments they have carried out which answer some of the possible objections of Goettel and Lewis—these additional experiments confirm their earlier results, ruling out the possibility of any significant amount of K in the core. Both sets of authors agree, however, that up to 1·5 per cent of the Earth's potassium may be in the core. This may just be sufficient to supply sufficient energy through the decay of K^{40} to set up convection in the core.

Seitz and Kushiro (1974) have carried out a series of experiments on the melting relations of a bulk sample of the Allende Type 3 carbonaceous chondrite to try and establish the initial evolutionary sequence of a planet of chondritic composition. Their experiments suggest that ferrobasaltic liquid can be generated by partial melting of materials similar in bulk composition to the Allende meteorite under anhydrous conditions and in the presence of less than 10 per cent metal. Under more reducing conditions, more metallic phase would precipitate, making the silicate melt more magnesian. The latter case would be expected in planets like the Earth, with a large metal core. If the core formed after accretion, it would be expected to contain a large proportion of the nickel and sulphur of the system. However in the experiments no potassium was detected in the sulphide phase.

References

Ahrens, T. J., Anderson, D. L. and Ringwood, A. E. (1969). Equations of state and crystal structures of high pressure phases of shocked silicates and oxides. *Rev. Geophys.* **4**, 667.

Al'tshuler, L. V., Krupnikov, K. K., Ledenev, B. N., Zhuckikhin, V. I. and Brazhnik, M. I. (1958a). Dynamic compressibility and equation of state of iron under high pressure. *Sov. Phys. JETP* **34**, 606.

Al'tshuler, L. V., Krupnikov, K. K. and Brazhnik, M. I. (1958b). Dynamic compressibility of metals under pressures from 400,000 to 4,000,000 atmospheres. *Sov. Phys. JETP* **34**, 614.

Al'tshuler, L. V., Bakanova, A. and Trunin, R. F. (1962). Shock adiabats and zero isotherms of seven metals at high pressure. *Sov. Phys. JETP* **15**, 65.

Anderson, D. L. and Hanks, T. C. (1972). Formation of the Earth's core. *Nature* **237**, 387.

Anderson, D. L., Sammis, C. and Jordan, T. (1971). Composition and evolution of the mantle and core. *Science* **171**, 1103.

Birch, F. (1963). Some geophysical applications of high pressure research. *In* "Solids Under Pressure" (Ed. W. Paul and D. M. Warschauer), McGraw-Hill, New York.

Birch, F. (1965). Energetics of core formation. *J. geophys. Res.* **70**, 6217.

Birch, F. (1968). On the possibility of large changes in the Earth's volume. *Phys. Earth Planet. Int.* **1**, 141.

Bolt, B. A. (1962). Gutenberg's early PKP observations. *Nature* **196**, 122.

Bolt, B. A. (1964). The velocity of seismic waves near the Earth's core. *Bull. seism. Soc. Am.* **54**, 191.

Bolt, B. A. (1972). The density distribution near the base of the mantle and near the Earth's centre. *Phys. Earth Planet. Int.* **5**, 301.

Brett, R. and Bell, P. M. (1969). Melting relations in the Fe-rich portion of the system Fe–FeS at 30 kb pressure. *Earth Planet. Sci. Letters* **6**, 479.

Bullen, K. E. (1946). A hypothesis of compressibility at pressures of the order of a million atmospheres. *Nature* **157**, 405.

Bullen, K. E. (1949). Compressibility-pressure hypothesis and the Earth's interior. *Mon. Not. R. astr. Soc. Geophys. Suppl.* **5**, 355.

Bullen, K. E. (1950). An Earth model based on a compressibility-pressure hypothesis. *Mon. Not. R. astr. Soc. Geophys. Suppl.* **6**, 50.

Bullen, K. E. (1963). An index of degree of chemical inhomogeneity in the Earth. *Geophys. J.* **7**, 584.

Bullen, K. E. (1965a). On compressibility and chemical inhomogeneity in the Earth's core. *Geophys. J.* **9**, 195.

Bullen, K. E. (1965b). Models for the density and elasticity of the Earth's lower core. *Geophys. J.* **9**, 233.

Bullen, K. E. (1969). Compressibility-pressure gradient and the constitution of the Earth's outer core. *Geophys. J.* **18**, 73.

Bullen, K. E. (1973a). Cores of the terrestrial planets.' *Nature* **243**, 68.

Bullen, K. E. (1973b). On planetary cores. *The Moon* **7**, 384.

Bullen, K. E. and Haddon, R. A. W. (1967). Earth models based on compressibility theory. *Phys. Earth Planet. Int.* **1**, 1.

Buolos, M. S. and Manuel, O. K. (1971). The Xenon record of extinct radio-activities in the Earth. *Science* **174**, 1334.

Clark, A. M. and Long, J. V. P. (1971). The anisotropic diffusion of nickel in olivine. *In* "Duffusion Processes" (Eds. J. N. Sherwood, A. V. Chadwick, W. M. Muir and F. L. Swinton), Gordon and Breach, New York.

Clark, S. P., Jr., Turekian, K. K. and Grossman, L. (1972). Model for the early history of the Earth. *In* "The Nature of the Solid Earth" (Ed. E. C. Robertson), McGraw-Hill, New York.

Cleary, J. R. (1974). The D'' region. *Phys. Earth Planet. Int.* **9**, 13.

Derr, J. (1969). Internal structure of the Earth inferred from free oscillations. *J. geophys. Res.* **74**, 5202.

Dziewonski, A. M. and Gilbert, F. (1972). Observations of normal modes from 84 recordings of the Alaska earthquake of 1964, March 28. *Geophys. J.* **27**, 393.

Flaser, F. M. and Birch, F. (1973). Energetics of core formation: a correction. *J. geophys. Res.* **78**, 6101.

Goettel, K. A. (1972). Partitioning of potassium between silicates and sulphide melts; experiments relevant to the Earth's core. *Phys. Earth Planet. Int.* **6**, 161.

Goettel, K. A. and Lewis, J. S. (1973). Comments on a paper by V. M. Oversby and A. E. Ringwood (Earth Planet. Sci. Letters 14, 345, 1972). *Earth Planet. Sci. Letters* **18**, 148.

Green, D. H. and Ringwood, A. E. (1963). Mineral assemblages in a model mantle composition. *J. geophys. Res.* **68**, 937.

Haddon, R. A. W. and Bullen, K. E. (1969). An Earth incorporating free Earth oscillation data. *Phys. Earth Planet. Int.* **2**, 35.

Hales, A. L. and Roberts, J. L. (1970). Shear velocities in the lower mantle and the radius of the core. *Bull. seism. Soc. Am.* **60**, 1427.

Hales, A. L. and Herrin, E. (1972). Travel-times of seismic waves. *In* "The Nature of the Solid Earth" (Ed. E. C. Robertson), McGraw-Hill, New York.

Hall, H. T. and Murthy, V. R. (1971). The early chemical history of the Earth; some critical elemental fractionations. *Earth Planet. Sci. Letters* **11**, 239.

Hanks, T. C. and Anderson, D. L. (1969). The early thermal history of the Earth. *Phys. Earth Planet. Int.* **2**, 19.

Higgins, G. and Kennedy, G. C. (1971). The adiabatic gradient and the melting point gradient in the core of the Earth. *J. geophys. Res.* **76**, 1870.

Jordan, T. H. (1972). Estimation of the radial variation of seismic velocities and density in the Earth. Ph.D. thesis, California Institute of Technology, 1972.

Jordan, T. H. and Anderson, D. L. (1974). Earth structure from free oscillations and travel-times. *Geophys. J.* **36**, 411.

King, D. A. and Ahrens, T. J. (1973). Shock compression of iron sulphide and the possible sulphur content of the Earth's core. *Nature* **243**, 82.

Knopoff, L. and Shapiro, J. N. (1969). Comments on the interrelations between Grüneisen's parameter and shock and isothermal equations of state. *J. geophys. Res.* **74**, 1439.

Kraut, E. A. and Kennedy, G. C. (1966). New melting law at high pressures. *Phys. Rev.* **151**, 668.

Lewis, J. S. (1971). Consequences of the presence of sulphur in the core of the Earth. *Earth Planet. Sci. Letters* **11**, 130.

Liu, Lin-Gun (1974). Birch's diagram: some new observations. *Phys. Earth Planet. Int.* **8**, 56.

Lyttleton, R. A. (1973). The end of the iron-core age. *The Moon* **7**, 422.

McQueen, R. G. and Marsh, S. P. (1960). Equation of state for nineteen metallic elements from shock-wave measurements to two megabars. *J. appl. Phys.* **31**, 1253.

McQueen, R. G. and Marsh, S. P. (1966). Shock wave compression of iron–nickel alloys and the Earth's core. *J. geophys. Res.* **71**, 1751.

Muirhead, K. J. and Cleary, J. R. (1969). Free oscillations of the Earth and the D'' layer. *Nature* **223**, 1146.

Murthy, V. R. and Hall, H. T. (1970). The chemical composition of the Earth's core; possibility of sulphur in the core. *Phys. Earth Planet. Int.* **2**, 276.

Murthy, V. R. and Hall, H. T. (1972). The origin and chemical composition of the Earth's core. *Phys. Earth Planet. Int.* **6**, 123.

Newton, R. C., Jayaraman, A. and Kennedy, G. C. (1962). The fusion curves of the alkali metals up to 50 kilobars. *J. geophys. Res.* **67**, 2559.

Oversby, V. M. and Ringwood, A. E. (1972). Potassium distribution between metal and silicate and its bearing on the occurrence of potassium in the Earth's core. *Earth Planet. Sci. Letters* **14**, 345.

Oversby, V. M. and Ringwood, A. E. (1973). Reply to comments by K. A. Goettel and J. S. Lewis. *Earth Planet. Sci. Letters* **18**, 151.

Press, F. (1968). Density distribution in the Earth. *Science* **160**, 1218.

Qamar, A. (1971). Seismic wave velocity in the Earth's core; a study of *PKP* and *PKKP*. Ph.D. Thesis, University of California, Berkeley, 1971.

Qamar, A. (1973). Revised velocities in the Earth's core. *Bull. seism. Soc. Am.* **63**, 1073.

Ramsey, W. H. (1949). On the nature of the Earth's core. *Mon. Not. R. astr. Soc. Geophys. Suppl.* **5**, 409.

Rice, M. H., McQueen, R. C. and Walsh, J. M. (1958). Compression of solids by strong shock waves. *In* "Solid State Physics" (Ed. F. Seitz and D. Turnbull), Vol. 6, Academic Press, New York.

Ringwood, A. E. (1959). On the chemical evolution and densities of the planets. *Geochim. cosmochim. Acta* **15**, 257.

Ringwood, A. E. (1961). Silicon in the metal phase of enstatite chondrites and some geochemical implications. *Geochim. cosmochim. Acta* **25**, 1.

Ringwood, A. E. (1966a). Chemical evolution of the terrestrial planets. *Geochim. cosmochim. Acta* **30**, 41.

Ringwood, A. E. (1966b). The chemical composition and origin of the Earth. *In* "Advances in Earth Science" (Ed. P. M. Hurley), M.I.T. Press, Cambridge, Mass.

Ryzhenko, B. and Kennedy, G. C. (1973). The effect of pressure on the eutectic in the system Fe–FeS. *Am. J. Sci.* **273**, 803.

Seitz, M. G. and Kushiro, I. (1974). Melting relations of the Allende meteorite. *Science* **183**, 954.

Shima, M. and Honda, M. (1967). Distribution of alkali, alkaline earth and rare elements in component minerals of chondrites. *Geochim. cosmochim. Acta* **31**, 1995.

Taggert, J. and Engdahl, E. R. (1968). Estimation of *PcP* travel-times and depth to the core. *Bull. seism. Soc. Am.* **58**, 1293.

Usselman, T. M. (1975a). Experimental approach to the state of the core: the liquidus relations of the Fe-rich portion of the Fe–Ni–S system from 30 to 100 kb. *Am. J. Sci.* (In press).

Usselman, T. M. (1975b). Experimental approach to the state of the core; composition and thermal regime. *Am. J. Sci.* (In press).

Van Thiel, M. (1967). Compendium of shock wave data. Univ. Calif., Radiation Lab. Rept. UCRL 50108, 1967.

Verhoogen, J. (1961). Heat balance of the Earth's core. *Geophys. J.* **4**, 278.

Walsh, J. M. and Christian, R. H. (1955). Equation of state of metals from shock wave measurements. *Phys. Rev.* **97**, 1544.

Walsh, J. M., Rice, M. H., McQueen, R. G. and Yarger, F. L. (1957). Shock-wave compressions of twenty-seven metals, equation of state of metals. *Phys. Rev.* **108**, 196.

Wasserburg, G. J., MacDonald, G. J. F., Hoyle, F. and Fowler, W. A. (1964). The relative contribution of uranium, thorium and potassium to heat production in the Earth. *Science* **143**, 465.

Wiik, H. B. (1956). The chemical composition of some stony meteorites. *Geochim. cosmochim. Acta* **9**, 279.

6. The Cores of Other Planets

6.1. Introduction

In this chapter the question of whether the other terrestrial planets (Venus, Mars and Mercury) and the moon have cores will be discussed. Since it is believed that the Earth's magnetic field arises from motions in the fluid, predominantly iron outer core, the question of whether these other bodies possess a magnetic field is of crucial importance to whether they also have cores. In this respect of all the known planetary bodies, only Jupiter and one of its satellites besides the Earth have been found to have a significant magnetic field. It must not be forgotten, however, that in the case of the Earth the poloidal field is probably only about one per cent of the toroidal field. Thus failure to observe a poloidal field in another planet does not by itself imply that the planet has no fluid core containing a toroidal field.

6.2. The Moon

In 1959 a magnetometer aboard Luna 2 passed within 55 km of the Moon's surface and detected no magnetic field. Taking into account the sensitivity of the magnetometer this put an upper limit of 6×10^{21} gauss cm³ on the moon's global magnetic moment—a value some four orders of magnitude lower than the present dipole moment of the Earth. This lack of any significant lunar magnetic field was confirmed later by Luna 10 (Dolginov et al., 1966) and Explorer 35 (Sonett et al., 1967) which showed that the upper limit of any lunar dipole moment was even lower being 6×10^{20} gauss cm³. A more detailed analysis of the Explorer 35 data carried out later by Behannon (1968) yielded a still lower value of 1×10^{20} gauss cm³, and the most recent estimate

213

of the maximum lunar dipole moment (based on Apollo subsatellite magneto-meter results) is 6×10^{19} gauss cm^3.

The results of the Apollo programme were very surprising. Stable compon-ents of natural remanent magnetization (NRM) of lunar origin were found in Apollo 11 samples by a number of workers (*Geochim. cosmochim. Acta* **34**, Suppl. 1, 1970, contains a number of papers on this topic). Later, lunar surface magnetometers at the Apollo 12, 14, 15 and 16 landing sites revealed surprisingly high local surface fields of tens and hundreds of γ (Sonett *et al.*, 1971). Finally magnetometer measurements made with Apollo 15 and 16 subsatellites orbiting the moon at a height $\simeq 110$ km detected sharp magnetic anomalies up to several γ on both the front and far side of the moon. These anomalies are apparently associated with topographic and/or geologic features, and imply a degree of homogeneous magnetization of the lunar crust on a scale of tens or hundreds of km (Coleman *et al.*, 1972).

The principal difference between lunar and terrestrial samples is that the remanent magnetization of the lunar samples is carried by iron with small amounts of nickel. The magnetic behaviour of these materials is very different from that of the familiar titanomagnetites and other ferromagnetics in terrestrial rocks. The interpretation of the NRM of the returned lunar samples ascribes it to three possible sources: magnetic contamination, secondary effects on the lunar surface, and primary magnetization associated with the formation of the rock. The primary NRM of igneous rocks is of the thermoremanent (TRM) type. In the case of the breccias the primary NRM will be thermoremanent if the Curie point is exceeded during formation but not if they have a relatively low temperature of origin. Among the secondary processes, impact-related shock is likely to be important. It can readily account for the observed inhomogeneity of some of the NRM, and clearly much of the lunar surface material has been shocked.

The origin of the NRM of the returned samples—particularly the origin of the fields in which the magnetization was acquired—is still unresolved. If it is necessary to invoke a lunar dynamo during the early history of the moon, this would place severe constraints on geochemical models of the moon's evolution. It would require that the moon differentiated early in its history in order to have a metallic fluid core. It is also not clear whether a single explana-tion will account for all aspects of the observed magnetic phenomena—the NRM of the lunar samples, the surface fields observed with lunar surface magnetometers and the fields observed by the subsatellites.

The NRM of the samples argues strongly for the existence of ancient lunar fields in which they acquired their magnetization. Unfortunately, there are no samples of demonstrably undisturbed bedrock, which can give the configura-tion of the magnetic fields. The stable magnetization carried by the rocks suggests that many of them were magnetized in fields of 10^3 or $10^4\gamma$. However, the range of fields implied by the natural remanence of the returned lunar rocks is remarkable, being from hundreds of γ to the order of an oersted.

Most suggestions appeal to a planetary wide lunar field that induced the observed magnetization of the rocks and which has subsequently disappeared. Possibilities include external fields, such as the solar wind or the terrestrial field, and fields of internal origin, such as a lunar dynamo. A difficulty with the proposal that the lunar magnetic field was acquired when the moon was much closer to the Earth, is that the range of ages of the magnetic lunar material returned to Earth is at least 500 m yr, i.e. the moon would have had to have been very close to the Earth for an improbably long time, and also at a distance close to the Roche limit.* There are also models that assume that the whole moon or a substantial part of it was magnetized early in its history giving rise to a primitive remanent field. Subsequently the rocks that are observed to be magnetized and those that give rise to present lunar remanent fields were magnetized in this primitive remanent field. Later still this field is postulated to have been thermally demagnetized; this hypothesis is consistent with some of the proposed thermal histories of the moon (e.g. Solomon and Toksöz, 1973). Such models may conveniently be referred to as fossil field models, since the magnetization that is now observed is held to have been acquired in the primitive remanent field of the moon and not directly in a solar wind field, a terrestrial field, or a lunar dynamo field.

The lunar dynamo theory has been most strongly advocated by Runcorn *et al.* (1971) and Strangway *et al.* (1971). It is held to have operated from some time before the oldest rocks studied were formed until some period after the last volcanic rocks and breccias that have been investigated, originated. At this point it then shut off. Much of the support for a lunar dynamo comes from the suggestion that lunar rocks for the most part exhibit NRM appropriate for TRM acquired in a field of about $10^3\gamma$. Fuller (1974) has pointed out however that this is not very well substantiated—it appears that it does not apply to certain of the rocks that have the most convincing TRM-like natural remanence. It may well be that the dynamo will have to account for a wide range of ancient lunar fields (including values as high as 1 Oe). Levy (1972) has also concluded that the moon would have to rotate faster than its strength would allow before it could give rise to a self-generating dynamo (his dynamo model is based on that of E. N. Parker—see Section 4.2).

Again an analysis of the available stable NRM data of lunar crystalline rocks and the more scanty paleointensity data of both crystalline rocks and breccias indicate no observable trend with age as might be expected from TRM due to a dipolar field from a central lunar core dynamo which is now dead. Furthermore, intensive paleointensity experiments on multiple samples

* At the Roche limit the self gravitational field of the moon would just balance the disruptive force of the Earth's gravitational field.

of similar age and collected from the same mission show the same type of widespread variation (10^3–$10^5\gamma$) as seen previously in representative samples of different ages.

Formation of a molten lunar core or pockets of pure Fe or Fe–Ni would require temperatures above the melting point of iron in the lunar interior. Extrapolating the data of Sterrett et al. (1965), the melting point of metallic iron at the centre of the moon is about 1660°C. The presence of Ni would lower this temperature only slightly. To produce a magnetic field 4000 m yr ago, such a core would have had to have been formed early. On the other hand temperatures of the order of 1660°C at the centre of the moon, early in lunar history, are unacceptable according to most current models of the moon's thermal history which call for a cool lunar interior (e.g. Hanks and Anderson, 1969; Wood, 1972; McConnell and Gast, 1972; Toksöz et al., 1972). Moreover unreasonably high accretional temperatures would be required to heat the centre of the moon to 1660°C in the first 500 m yr of its history.

The presence of S in the core of the moon would allow a molten lunar core at considerably lower temperatures (~ 1000°C). The eutectic temperature of 988°C (at 1 atm)* would also be lowered somewhat by the presence of Ni. Anderson (1972) has calculated that the density of an Fe–FeS eutectic mix at 30 kbar pressure is about 5·3 g/cm^3. This value is significantly less than the density of pure Fe which Runcorn et al. (1970) assumed in their calculations on the maximum permissible size of a lunar core. A core of lower density than Fe could be larger than one of pure Fe without violating the constraints imposed by the moon's moments of inertia.

The moon does not have an internally generated magnetic field today, so that, if the field was caused by a convecting core, either the core froze or the magnetic Reynolds number fell below the critical value (Runcorn, 1972). Failure of the lunar dynamo by solidification of a core or of deep pockets of Fe–FeS is inconsistent with most models of lunar thermal history, nearly all of which require the temperature of the deep interior to be considerably above 1200°C today. On the other hand such thermal models are consistent with the present model for a core provided that the lunar dynamo ceased because the magnetic Reynolds number fell below the critical value. If the dynamo failed by solidification, it is still possible to modify most thermal models by postulating convective cooling. (Tozer (1972) believes that the moon is likely to undergo solid state convective motion as soon as the temperature of the lunar interior rises above about 1000°C, and that the central temperature adjusts to a steady state temperature of between 600 and

---

* Usselman (1972) has extended the earlier work of Brett and Bell (1969) on the effect of pressure on the Fe–FeS eutectic temperature. At a pressure of 50 kbar (the approximate pressure at the centre of the moon), the eutectic temperature is ~ 1000°C.

1000°C.) Convection would also aid in core formation. If the dynamo failed by solidification and the dynamo was a central core rather than a pocket or a number of pockets, it is unlikely that temperatures within the moon could have exceeded the solidification temperature of the core at any time afterwards. Brett (1973) has re-examined the geochemical arguments in favour of a lunar dynamo and produced an *ad hoc* model which attempts to reconcile, as much as possible, some of these divergent opinions.

Solomon and Toksöz (1973) have also investigated the internal constitution

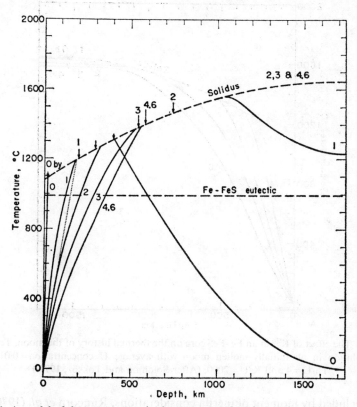

Fig. 6.1. A model of the temperature in the moon as a function of time since lunar origin. Time, in 1000 m yr, is indicated by the number adjacent to each profile. The initial temperature distribution is calculated from a simple model of the accretion process (Toksöz *et al.*, 1972); accretion at 0°C over a 100-yr time span is assumed. The moon is partially or completely molten at those depths where the temperature profile lies along the solidus of mare basalt (Ringwood and Essene, 1970). The depth range for complete melting is delimited by the small arrows above the solidus. Also shown are the Fe–FeS eutectic temperature (Brett and Bell, 1969) and the phase boundary between feldspathic and spinel pyroxenite (dotted line) in the model lunar mantle composition of Ringwood and Essene (1970). (After Soloman and Toksöz, 1973.)

H

and evolution of the moon. Temperatures 500 m yr after lunar origin exceed the Fe–FeS eutectic, in their model shown in Fig. 6.1, throughout most of the moon. Thus core formation by melting, sinking and aggregation of iron-rich droplets is achieved prior to the formation of the oldest lunar rocks possessing remanent magnetization. Thermal history models with initially cold interiors depleted in radioactive elements, are inconsistent with an early molten core. Solomon and Toksöz also showed that the presence today of a pure-Fe core is

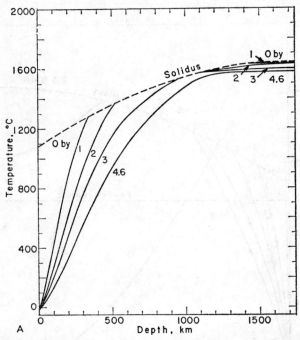

Fig. 6.2. The effect of K^{40} in an Fe–FeS core on the thermal history of the moon. Temperature evolution in an initially molten moon with average U concentration $= 0.011$ ppm, Th/U $= 4$ and K/U $= 2000$. (After Solomon and Toksöz, 1973.)

not precluded by moment of inertia considerations. Runcorn *et al.* (1970) had earlier shown that a core would not significantly change the moon's moments of inertia provided that it were less than about 20 per cent of the lunar radius in size.

If the moon contains a small iron-rich core with additional sulphur, then the suggestion of Lewis (1971) and Hall and Murthy (1971) that potassium would be enriched in an Fe–FeS melt relative to silicates has important implications for its thermal evolution. The thermal evolution for an initially molten moon with U $= 0.011$ ppm, K/U $= 2000$, and Th/U $= 4$ is shown in

Fig. 6.2. All radioactive elements are concentrated toward the lunar surface shortly after the time of lunar formation. The model predicts a solid moon today. Figure. 6.3 shows the thermal evolution of a moon with the same initial temperature profile and U and Th abundances, but with a chondritic K/U ratio of 80,000 (Wasserburg et al., 1964). The near surface radioactive heat generation is the same as in Fig. 6.2, but 39/40 of the potassium is assumed to remain in a 340-km radius core. The result, is spectacular. Heat from the

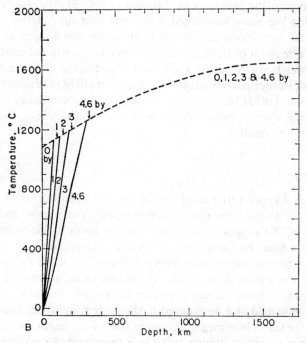

Fig. 6.3. Thermal evolution for a moon similar to that in Fig. 6.2 except that K/U = 80,000. (After Solomon and Toksöz, 1973.)

additional potassium is sufficient to keep most of the moon melted throughout its history—the moon is molten below 300 km at present.

If we assume that the moon accreted cold but was heated from the outside during the final stages of accretion, this would lead to a molten shell very early on in lunar history, which later formed the crust as indicated by seismic studies. The interior of the moon, however, would have been cool during its early history and would have warmed up because of radioactive heating only relatively recently, thus forming a molten or partially molten core which is also indicated by seismic data (Nakamura et al., 1973). Strangway and Sharpe

(1974) have recently considered whether an early hot outside, cold inside moon, which evolved into the present cold outside, hot inside moon, could quantitatively account for the evidence of an ancient magnetic field. Two questions need to be answered. First, what is the origin of the magnetizing field? Secondly, once the moon had become magnetized by an isothermal process could the memory have been retained until the termination of igneous activity about 3200 m yr ago? There is no definitive answer to the first question, because there are no clear records of planetary fields predating those frozen into the lunar rocks. Strangway and Sharpe showed however that provided the moon contained a few per cent of metallic Fe and was exposed to an extra-lunar field of about 10–20 Oe while much of it was still below the Curie point of Fe, it is possible to derive a restricted class of thermal evolution models which satisfy the known constraints. The question of the time decay of isothermal remanent magnetization (IRM) is difficult to answer, but Gose et al. (1972) have shown, by extrapolating the decay rate of IRM acquired in the laboratory in multi-domain, iron-bearing rocks, that the decay coefficient is very small.

6.3. Mars

The Mariner 4 spacecraft passed within about 13,200 km of Mars on July 14–15th, 1965. During the close encounter with the planet, measurements were made of the magnetic field and various particle fluxes, all of which indicated that Mars had at most a very weak magnetic field.

We do not have a quantitative theory of the origin of the Earth's radiation belts and it is not possible to predict the nature of such belts for a planet at arbitrary distance from the sun: however the planet must have a sufficiently strong magnetic field and be exposed to the solar wind. Since the distance of Mars from the sun is intermediate between that of the Earth and Jupiter, both of which have intense radiation belts, it is reasonable to assume that Mars also would have radiation belts provided that it is a sufficient magnetized body. A system of sensitive particle detectors on Mariner 4 indicated the presence of electrons of energy > 40 kev out to a radial distance of 165,000 km on the morning side of the Earth, yet failed to detect any such electrons during the close encounter with Mars (Van Allen et al., 1965). This implies that the magnetic dipole moment of Mars is less than 0·001 that of the Earth, i.e. the upper limit on the equatorial magnetic field at the surface of Mars is about 200γ.

Similar results were reported by O'Gallagher and Simpson (1965). The Mariner 4 carried a solid-state charged particle telescope capable of detecting electrons with energies greater than 40 kev and protons with energies greater than 1 Mev. The trajectory of Mariner 4 would have carried it through a bow shock, transition region and magnetospheric boundary had these existed. No evidence of charged particle radiation was found in any of these regions. Again a planet with even a very small magnetic field might be expected to produce a wake in the anti-solar direction. Mariner 4 passed sufficiently close to Mars to have detected such a wake, had it existed—no escape of electrons, as would be expected in such a wake, was observed. O'Gallagher and Simpson placed the same upper limit on any Martian magnetic field, viz. 0·1 per cent of that of the Earth.

Mariner 4 also carried a magnetometer during the close encounter with Mars; no magnetic effects were observed that could be definitely associated with a Martian magnetic field. Smith *et al.* (1965a) put an upper limit for a Martian magnetic moment of 3×10^{-4} that of the Earth.

Dolginov *et al.* (1972, 1973), in an analysis of the magnetograms transmitted from the USSR Mars 2 and 3 orbiting probes, claimed that Mars possesses an internal magnetic field. The magnetograms indicated a field intensity some seven to ten times greater than that of the interplanetary field at the distance of the orbit of Mars. They concluded that Mars possesses a dipole moment of $\sim 2·4 \times 10^{22}$ gauss cm³ and an intensity of $\sim 60\gamma$ at the magnetic equator. These values are still less than the upper limit deduced from the analysis or the Mariner 4 data. It must also be pointed out that Bogdanov and Vaisberg (1974—preprint) have interpreted measurements of ion fluxes on the Mars 2 and 3 space craft differently. They suggest that the observed magnetic field variations are a consequence of a solar wind—planetary ionosphere interaction rather than that of an internal planetary dipole magnetic field.

Since the rotational period of Mars is approximately the same as that of the Earth, its much weaker magnetic moment would suggest that it has at most a very small fluid, electrically conducting core. The density and moment of inertia coefficient of Mars show that it possesses a ratio of iron to silicate smaller than that of the Earth but larger than that of the moon. The radius of an iron core should not exceed a few hundred km. Such a core could have sustained, through dynamo action, a stronger magnetic field in earlier times than any now present. Runcorn (1972) is of the opinion that the field of 60γ, of apparent internal origin, detected by the Mars 2 orbiter is more likely to be due to such a dynamo now acting weakly than to permanent magnetization, though the latter may be present in the crustal rocks as it is on the Moon.

On Bullen's Fe_2O hypothesis (see Section 5.5), Mars would have an iron core of radius ~ 1400 km, but no Fe_2O zone and therefore presumably no fluid zone. This would be compatible with the failure to observe any significant magnetic field. Lewis (1972) concluded that Mars is essentially devoid of free iron. It may possess a small core of FeS with or without a small amount of

Fe^0—its mantle is rich in FeO, with a FeO/FeO+MgO ratio $\simeq 0.5$. His conclusions are based on his model of an initial nebula of solar composition in which there is a steep radial temperature gradient.

Studies of the internal structure of Mars have been hampered by uncertainty in its radius, values of which have ranged from 3310 to 3423 km. However an occultation experiment on board Mariner 4 led to a very accurate value (Fjeldbo, et al., 1966), the accepted figure for the equatorial radius now being 3394 ± 5 km. This figure gives for the moment of inertia coefficient z the value 0·377. Binder (1969) has constructed a number of models of the internal structure of Mars using the Mariner 4 data and recent investigations (Anderson, 1967) on the composition and structure of the Earth's upper mantle, assuming that there is no radical difference between mantle materials of Mars and mantle materials of the Earth. He found that Mars has an iron core with a radius in the range 790–950 km and a mass in the range 2·7–4·9 per cent of the total mass of the planet. The percentage of free iron in Mars is considerably lower than that in the Earth whose core contains about 31 per cent of the Earth's mass. Binder also concluded that in its early history the iron core of Mars was molten, adding support to his earlier (1966) suggestion that initially the internal conditions of Mars satisfied the requirements of dynamo theory, Mars possessing an early magnetic field. Models (e.g. Anderson and Kovach, 1967) in which Mars has a smaller, less dense core than the Earth with a mantle more dense than the Earth's have often been interpreted as suggesting that Mars has not differentiated completely, so that both core and mantle are cross-contaminated with each other's material. The data can however be simply explained if the core of Mars is pure FeS and the mantle very FeO-rich silicates as suggested by Lewis (1972).

A number of years ago, Ringwood (1959, 1966) proposed an alternative model for the constitution of Mars, viz. that the relative abundances of the common metals Fe, Mg, Si, Ca, Al, etc. in Mars are similar to those in the Earth, Venus, the Sun and Type 1 carbonaceous chondrites and that the differing densities between the planets are caused primarily by the varying amounts of oxygen present. Mars is believed to be completely oxidized, containing no free Fe. In a later paper, Ringwood and Clark (1971) investigated this hypothesis in some detail—this was made possible by the more recent accurate determinations of the mean density and moment of inertia of Mars, together with advances in high pressure research enabling the role of phase transitions within Mars to be evaluated. In their preferred model (see Fig. 6.4), magnetite has segregated to form a core with a radius of 1638 km (21 per cent of the mass of Mars)—the magnetite being present as the high pressure polymorph. If differentiation has proceeded sufficiently to form a core, it is likely that a crust is also present. It is also likely that a small metallic Fe core would be possible within the constraints laid down by their models. However an Fe core is not readily reconcilable with the inferred oxidized state of surface rocks and atmospheric composition.

Ringwood and Clark's final model is also in agreement with the probable thermal history of Mars. (MacDonald (1962) had shown that if Mars has the chondritic abundances of U, Th and K, the radioactive heat generated over 4500 m yr would have caused extensive internal melting and differentiation.) The absence of a large, electrically conducting convective liquid core in their model is also consistent with the failure to observe a significant magnetic field on Mars.

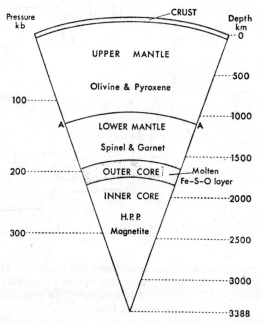

Fig. 6.4. Hypothetical cross-section through Mars. (After Ringwood and Clark, 1971.)

Anderson (1972) has also constructed a number of models of the possible internal structure of Mars (see Fig. 6.5). He finds that Mars cannot be homogeneous but must have a core, the size of the core depending on its density. Taking the density of pure Fe and pure troilite (FeS) as upper and lower bounds for the density of the core, he found that its radius must lie between 0·36 and 0·60 of that of Mars (i.e. between about 1220 and 2035 km). A meteoritic model for Mars leads to an Fe–S–Ni core, of 12 per cent mass of the planet. Mars has an Fe content of 25 wt per cent which is significantly less than that of the other terrestrial planets, but is close to the total Fe content of ordinary and carbonaceous chondrites. Anderson obtained a satisfactory model for Mars on the assumption that ordinary chondritic material was subjected to relatively moderate temperatures. Core formation would begin when temperatures exceeded the eutectic temperature in the system Fe–FeS (~990°C), but would not be completed unless temperatures exceeded the liquidus throughout most of the planet. Anderson estimated that approximately 63 per cent of the potential core-forming material (Fe–S–Ni) has entered the core. Thus Mars, in contrast to the Earth, is an incompletely differentiated planet, and its core is considerably richer in S than is the Earth's core. The absence of a significant magnetic field may be due to the smaller size of its

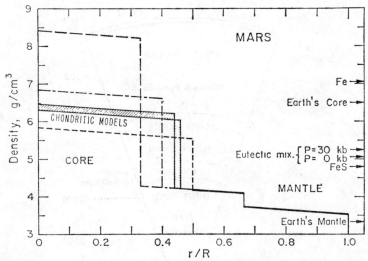

Fig. 6.5. Density versus radius for some representative models of Mars that have the correct mean density and moment of inertia. Models with chondritic compositions fall within the hatched region. All models have approximately the same mantle density. (After Anderson, 1972.)

core, as compared with that of the Earth, and its higher resistivity (due to large amounts of sulphur in the core). Alternatively if the dynamo is driven by precessional torques, the absence of a magnetic field may be due to the lack of a significant lunar torque.

In view of Anderson's (1972) study and the new data returned by Mariner 6, 7 and 9, Binder and Davis (1973) constructed a number of new models of the internal structure of Mars and reassessed the conclusions of the earlier work of Binder (1969). The mean equatorial radius of Mars as determined from the Mariner 4, 6 and 7 missions and ground-based radar observations is 3394 km. This radius assumes that the surface of Mars is an oblate spheroid whose flattening is equal to the dynamical flattening 1/190, i.e. Mars is in complete hydrostatic equilibrium. However, preliminary analysis of the Mariner 9 occultation data (Cain *et al.*, 1972) indicates that the surface of Mars may best be represented by a triaxial ellipsoid with equatorial radii of 3401 and 3395 km and a polar radius of 3372 km. The radius of a sphere whose volume is equal to that of this ellipsoid is 3389 km. In addition the actual planetary figure is extremely important in interpreting measurements of the coefficient J_2 of the second zonal harmonic in the aeropotential field in terms of the moment of inertia. If the planet is in perfect hydrostatic equilibrium, the measured value of J_2 is directly related to the moment of inertia through the Darwin–Radau relationship. If the planet is not in hydrostatic equilibrium, the computation of the moment of inertia coefficient z about the polar axis from the measured value of J_2 depends on the model used to explain the deviation from hydrostatic equilibrium. In addition to the case of hydrostatic equilibrium, Binder and Davis considered the case where Mars has a crust which is in isostatic equilibrium varying in thickness from the equator to the poles. They also considered the case in which deviations from hydrostatic equilibrium are supported by the mechanical strength of the crust. They found that z depends quite critically on crustal structure, so that it is not possible to accurately determine the internal structure of Mars. Calculations based on complete hydrostatic equilibrium give only a lower limit to the size of the core ~ 3–5 per cent of the planet's mass. Reasonable estimates of crustal structure indicate that Mars has a core radius of approximately 1250 km and a mass 9–10 per cent of that of the planet. The lack of exact knowledge of the polar and equatorial thicknesses of the crust does not significantly affect these results if the crust is in isostatic equilibrium. However the fact that at least some of the equatorial bulge may be uncompensated, makes it possible to construct reasonable models in which the core contains up to 25 per cent of the mass of the planet, i.e. almost as much of the planet's mass as does the core of the Earth.

Young and Schubert (1974) have re-examined the thermal history of Mars. Photographs obtained by Mariner 9 have revealed evidence of extensive volcanism (including a series of prominent shield volcanoes), mildly cratered plains similar to the lunar maria, and circular features such as domes and craters. This suggests that at some time in the past Mars must have been hot enough for partial melting to have occurred at depth, but gives no indication of present conditions.

In contrast to earlier studies, Young and Schubert considered solid state convection as a significant means of heat transport. They investigated two models—one with an iron core (radius and mass 0·4 and 0·1 that of Mars) and the other with a pure FeS core (radius and mass 0·6 and 0·26 that of Mars). These core models are the maximum density-minimum size and maximum size-minimum density cores consistent with the observed mass, radius and moment of inertia of Mars. The radioactive heat source concentration in the crust-mantle was taken to be the same as that in the terrestrial mantle ($2 \cdot 61 \times 10^{-7}$ erg cm^{-3} sec). Young and Schubert found convection in the mantle to be very effective in keeping temperatures relatively low. Thus for a pure iron core a temperature difference of 8000°C between the MCB and the surface when there is no convection is reduced to about 2300°C when convection occurs at a Raleigh number 100 times the critical value for the onset of convection. (The actual Rayleigh number is uncertain because viscosities are not known.) It is clear from their work, however, that for pure conduction, temperatures exceed typical silicate melting temperatures implying that convection in the Martian mantle is actually necessary to prevent large-scale melting there. Young and Schubert also found that whether Mars has a liquid or solid core depends on the effective viscosity ν of its mantle at temperatures above about 1500°C. If $\nu \lesssim 10^{22}$–10^{23} cm^2/sec at such temperatures, and evidence suggests that this is probably the case, then convection is so efficient as to cool nearly the entire mantle, and in particular the MCB, to temperatures below those at which Fe or Fe–FeS cores would be liquid. If such were the case, on the assumption that a planetary magnetic field could only originate through dynamo action in a liquid core, Mars would not have an intrinsic magnetic field.

Johnston et al. (1974) have also re-examined the thermal history and evolution of Mars, and come to somewhat different conclusions. They obtained a number of models of the Martian interior which had to satisfy certain constraints (mean density, moment of inertia and recent volcanism as inferred from the Mariner 9 photographs). Assuming an Fe–FeS core, they found that core formation and differentiation of a crust took place within the first 1000 m yr of the planet's history. At the present time the radius of the core is between about 1300 and 1800 km depending upon the exact composition, and the core is liquid even if its composition is not that of the eutectic.

6.4. Venus

The first measurements of the environment of Venus were made by the spacecraft Mariner 2, which on December 14th, 1962, passed within about 41,000 km of the centre of the planet. No disturbances of the interplanetary medium which could be attributed to the planet were observed (Frank et al., 1963; Smith et al., 1963). A search for charged particles magnetically trapped around Venus and measurements made by the magnetometer carried by Mariner 2 gave no indication that Venus possesses a magnetic field: an upper limit to the magnetic dipole moment of Venus of about 1/10–1/20 that of the Earth was deduced.

Later, Venera 4 (Dolginov et al., 1968, 1969; Gringauz et al., 1968) and

Mariner 5 (Bridge *et al.*, 1967) carried out magnetic field and plasma measurements one day apart in October 1967. When Venera 4 was 200 km above the surface of Venus it indicated a magnetic moment $< 10^{-4}$ that of the Earth (equivalent to a surface equatorial field strength $< 4\gamma$). Such a low value would be insufficient to deflect the flow of the solar wind as does the Earth's magnetic field. Mariner 5 passed about 10,150 km from the centre of Venus on October 19, 1967 and observed abrupt changes in the amplitude of magnetic fluctuations, in field strength and in plasma properties. Data from both the magnetometer and plasma probe showed clear evidence for the existence of a bow shock around Venus, similar to, but much smaller than, that near the Earth. The solar wind appears to flow around the planet without striking it, in contrast to the case of the moon: it appears to be deflected by a dense ionosphere on the sunlit side of Venus. Its high electrical conductivity prevents the passage of the incoming magnetic field, the "pile-up" of the field altering the plasma flow. Strong support for this interpretation comes from the observation of the Mariner Stanford Group (1967) that the upper boundary of the daylight ionosphere is very sharp and is pushed down to within about 500 km of the planet by the momentum of the solar wind. In the case of the moon the plasma ions are absorbed by the lunar surface and no shock develops (Lyon *et al.*, 1967): the moon appears to be a sufficiently good insulator to allow the interplanetary field to be convected through it essentially unaltered (Colburn *et al.*, 1967; Ness *et al.*, 1967). Although the shock around Venus resembles that around the Earth (except in scale) conditions inside it are quite different —in the case of Venus it is the ionosphere that deflects the solar wind, in the case of the Earth, it is its magnetic field.

The above observations also place an upper bound on the magnetic dipole moment of Venus: it cannot exceed 0·001 that of the Earth (within a factor of two). Similar conclusions were drawn by Van Allen *et al.* (1967) from the absence of energetic electrons (> 45 kev) and protons (> 320 kev) during the flyby of Mariner 5—they concluded that the dipole moment of Venus is almost certainly less than 0·01 and probably less than 0·001 that of the Earth. The next spacecraft to study the plasma field environment of Venus were Venera 6 which impacted the planet in May 1969 (Gringauz *et al.*, 1970) and Mariner 10 (Ness *et al.*, 1974a) which passed within 11,900 km from the centre of Venus on February 5, 1974. Because of the trajectory of Mariner 10, data obtained from its dual magnetometer system cannot reduce further the upper limit to the magnetic moment of Venus obtained by Venera 5 (10^{-4} that of the Earth's).

Because of its dense cloud cover, no reliable estimates of the rotation rate of Venus were possible prior to radar measurements in 1962 (Carpenter, 1964; Goldstein, 1964). These led to the surprising result that the rotation rate is slow (~ 250 days) and retrograde. More recent determinations indicate the period to be within a day or two of 243·15 days for which period Venus would present the same face to the Earth at every inferior conjunction—to an observer on Earth Venus would appear to rotate exactly 4 times between close approaches. The spin of Venus may thus be controlled by the Earth. The most recent and accurate determination of the rotation period (242·98 days) by Carpenter (1970) from radar tracking of features and Doppler broadening differs significantly from the Earth's synchronous value, and it is now very uncertain whether Venus is, or is not, in synodic spin resonance.

Goldreich and Peale (1970) have investigated the dynamical implications of the anomalous obliquity* of Venus. They found that solar gravitational tides acting on Venus tend to reduce the planet's obliquity to a value less than 90°— this result is true whether or not Venus is locked in synodic spin resonance with the Earth. They discussed two possible mechanisms which may stabilize the obliquity near 180°—atmospheric torques due to thermal tides and frictional dissipation of energy at the boundary between a rigid mantle and a differentially rotating liquid core. The atmospheric torque model is speculative because of large uncertainties in the properties of the Venusian atmosphere and Goldreich and Peale thus prefer the second possibility. Core-mantle interaction, which is induced by precession, must produce a reduction of the component of angular momentum in the orbital plane. Goldreich and Peale showed that such a mechanism is capable of driving the obliquity to 180° from values greater than 90° for a wide range of reasonable core viscosities and spin angular velocities. They had previously (1967) invoked coupling between a rigid mantle and liquid core as the means of capturing Venus into synodic spin resonance with the Earth. Thus both the surprising and un-expected features of the rotation of Venus—the possibility of synodic spin resonance with the Earth and obliquity close to 180°—can be accounted for if it possesses a fluid core similar to that of the Earth.

Since Venus has about the same size and average density as the Earth, it is quite probable that it does possess a metallic core with internal temperatures similar to those in the Earth. In such a case the very small (perhaps zero) value of its intrinsic magnetic moment may be due to its very slow rate of rotation with the consequent weakness of dynamo EM forces. On Bullen's Fe_2O hypothesis (see Section 5.4), the outer core of Venus is only about 600 km in thickness (Bullen, 1973)—Bullen suggested that allowance for compressibility may reduce this estimate still further and that this may be the reason for the failure to detect any significant magnetic field. Lewis (1972) concluded that Venus has a massive core of Fe–Ni alloy surmounted by a mantle of Fe^{2+}-free magnesium silicates. His conclusions are based on his

---

* A planet's obliquity is defined as the angle between its spin and orbital angular momenta. Uranus and Venus are anomalous among solar system bodies in that their spin angular momenta have components which are antiparallel to their orbital angular momenta. The obliquity of Uranus is 97° and that of Venus close to 180°.

assumption that the bulk composition of condensates in the solar nebula is determined by chemical equilibrium between the condensates and gases in a system of solar composition.

6.5. Mercury

Because of its small size, proximity to the sun and low reflectivity, it is very difficult to observe telescopically and photograph the markings on Mercury. However since the 1880's most astronomers believed that Mercury was rotating slowly with a period equal to its orbital period of 88 days. It was not until 1965 (Pettengill and Dyce) that delay Doppler maps of the surface with radar showed the rotation period to be 59 ± 3 days. Following these radar observations, several groups have independently re-examined the visual determinations of the rotation period and concluded that the optical data are consistent with a rotation period of about 59 days—in fact the most accurate rotation rate for Mercury (58·66 days) has been obtained from visual data (Smith and Rees, 1968; Chapman, 1968). Colombo (1965) first noticed that the observed sidereal spin period T_s was nearly two-thirds of the 88 day orbital period T_0 and suggested that the axial rotation might be "locked" to the orbital motion in a three-halves resonance state by the additional solar torque exerted on an axial asymmetry in Mercury's inertia ellipsoid. Colombo and Shapiro (1966) investigated a two-dimensional model of Sun-Mercury interaction and showed that resonances occur when a planet makes an integral number of half rotations during one orbital revolution, i.e. when

$$T_s = 2T_0/k \quad (k \text{ an integer}) \tag{3.1}$$

Thus the solar torque exerted on the permanent asymmetry could cause Mercury to be trapped into a $k=3$ resonance spin state (see also Goldreich and Peale, 1966).

Little work has been done on the internal constitution of Mercury. Constraints on models of the interior of a planet are given by astronomical data such as its mass and mean diameter. Recent determinations of these parameters by radar techniques have a much greater accuracy than that attainable by ordinary optical means. The mean density of Mercury is about the same as that of the Earth and the problem is to account for this for a small planet with a mass intermediate between that of Mars and the Moon. If the composition of Mercury is anything like that of the other terrestrial planets, it must have a very large iron core surrounded by a thin rocky mantle. Plagemann (1965) estimated the core radius to be about 0·86 that of the planet. He also estimated the temperature distribution in the planet and concluded that neither the core nor the mantle is molten. This together with its slow rate of rotation would seem to prohibit any planetary magnetic field by dynamo action. However his thermal calculations indicate that inside Mercury, heat flows continuously through the MCB, which could be considered a natural semi-conducting junction. Potential differences may then be generated by the Seebeck effect. If a complete circuit exists, it is possible that a small magnetic field may be set up. Lewis (1972) in a discussion of the chemistry of the solar system based on the sequence of condensations in the cooling solar nebula concluded that Mercury has a massive core of Fe–Ni alloy surmounted by a small mantle of Fe^{2+} free magnesium silicates. The high density of Mercury is thus attributed to accretion at temperatures so high that $MgSiO_3$ is only partially retained but Fe metal is condensed.

More recently Siegfried II and Solomon (1974) have carried out a series of calculations on the thermal evolution of Mercury using the gross physical properties of the planet and the cosmochemical model of Lewis (1972b). Their computations are based on the heat conduction equation. They use the finite difference method described by Toksöz et al. (1972) which allows for melting, differentiation and simulated convection. When the melting temperature of Fe is reached at a grid point, and an additional amount of heat corresponding to the heat of fusion of iron has been produced, the material at that point is considered to have melted. Within the melted region, complete differentiation of silicate from metal is assumed. The lower grid points in the region are assigned physical property values corresponding to iron, and the upper grid points values appropriate to a silicate. All radioactive heat sources are assumed to go into the silicate portion, since U and Th can be accommodated in silicate structures fairly readily. Convection is simulated in the molten silicate layer by holding the temperature to the melting curve and shifting any excess heat upwards. Solid state convection is not considered, although it may dominate heat transfer in most of the Earth's mantle at present. Siegfried and Solomon argue that because of the high thermal conductivity of the metallic fraction of Mercury, solid state convection is unlikely to postpone or prevent differentiation of an originally homogeneous planet—its principal effect would probably be, as in the case of the moon, to lower temperatures in the mantle by a few hundred degrees for some time after differentiation until the present.

Siegfried and Solomon's calculations indicate that once the temperature at any depth reaches the melting point, differentiation proceeds fully within approximately 500 m yr. This rapid segregation of a metal core is accompanied by extensive melting in the silicate mantle and might give rise to surface volcanism. The present day temperature profiles for all models involving differentiation are very similar, exhibiting a rise from surface temperature to the melting curve in the upper 300 km, a molten convecting mantle, and a slowly cooling core with temperature increasing with depth, either on or below the melting curve depending on the length of time since differentiation.

Figures 6.6 and 6.7 show representative thermal histories for models with and without differentiation.

The calculations of Lewis (1972) imply that the heavy elements will condense completely at the temperature expected near Mercury's orbit, so that the U/Fe and Th/Fe ratios in Mercury should reflect the cosmic abundances of these elements. Such a high concentration of radioactive heat sources implies

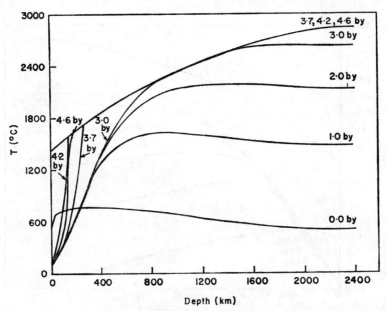

Fig. 6.6. A model for the thermal evolution of Mercury in which differentiation of core from mantle has occurred. The initial temperature profile is derived from a Hanks–Anderson (1969) accretion rate, a total accretion time of 10^5 yr, and an accretion temperature of 750°K. The average present-day U abundance is 0.44×10^{-7} g/g. The thermal conductivity of the metal-silicate mix prior to differentiation is the arithmetic mean of the Hashin–Shtrikman (1963) bounds. The adopted solidus is shown at the top of the figure. The time since planetary origin in 1000 m yr is shown adjacent to the corresponding temperature profile. (After Siegfried II and Solomon, 1974.)

differentiation of the planet into mantle and core for almost any probable initial temperature profile. Once the planet has differentiated, it quickly reaches almost a steady state, where radioactive heating keeps a large portion of the silicate mantle partially molten and convecting, and the iron core cools slowly. As long as differentiation occurred, the present thermal state is not very sensitive to initial conditions.

Figure 6.8 (after Siegfried and Solomon) shows the density and pressure distributions for a fully differentiated model based on the thermal history of

Fig. 6.7 and for a cooler, undifferentiated model based on the thermal evolution of Fig. 6.6.

Information about the magnetic field environment of Mercury has been obtained from two triaxial fluxgate magnetometers carried by Mariner 10 which flew by the planet on March 29, 1974 (Ness *et al.*, 1974b). Unexpectedly a very well developed, detached bow shock wave was observed. In addition a magnetosphere-like region with a maximum field strength of 98γ was observed

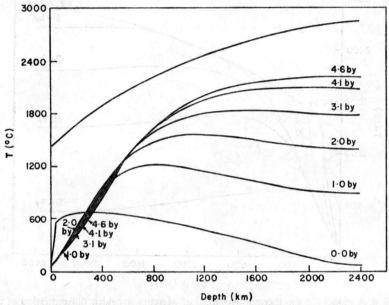

Fig. 6.7. A model for the thermal evolution of Mercury. The initial temperature profile is derived from a Hanks–Anderson (1969) accretion rate, a total accretion time of 10^5 yr, and an accretion temperature of 350°K. The average present-day U abundance is $0\cdot304 \times 10^{-7}$ g/g. The thermal conductivity is the upper Hashin–Shtrikman (1963) bound on the iron-silicate mixture. Solidus and labelling of temperature profiles are as in Fig. 6.6. No differentiation has occurred in this model. (After Siegfried II and Solomon, 1974.)

at closest approach (704 km above the planetary surface) with boundaries similar to the terrestrial magnetopause. This is a factor of 5 greater than the average interplanetary magnetic field strength (18γ) measured outside the bow shock.

It is not possible to determine uniquely the origin of the enhanced magnetic field. It may be intrinsic to the planet, distorted by interaction with the solar wind. Alternatively it may be associated with a complex induction process whereby the planetary interior—atmosphere—ionosphere interact with the solar wind to generate, by dynamo action, the observed field. Ness *et al.*

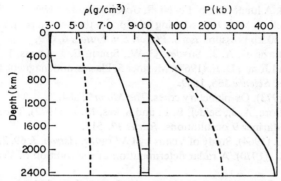

Fig. 6.8. Density and pressure profiles for two models of the internal structure of Mercury. The solid lines represent a fully differentiated model based on the thermal history of Fig. 6.6. The dashed lines represent a cooler, undifferentiated model based on the thermal evolution of Fig. 6.7. (After Siegfried II and Solomon, 1974.)

believe that the balance of evidence favours the conclusion that Mercury possesses an intrinsic magnetic field, and calculated the coordinates of a simple model of an offset, tilted dipole. This dipole is oriented within 20° of the ecliptic pole, i.e. it is almost aligned with the axis of rotation of the planet. The dipole is offset $0.47\ R_m$—considering the very large size of the core of Mercury (due to its high mean density), such a large offset is not implausible. The presence of an intrinsic planetary magnetic field may be due to a dynamo currently active within the planetary interior or a residual remanent magnetic field associated with a now extinct dynamo. Thus it is possible that Mercury rotated faster earlier in its history. Ness *et al.* stress the preliminary nature of their analysis, and point out that temporal variations of the structure of the magnetosphere of Mercury would masquerade as spatial variations of the magnetic field in the interpretation of data from a single flyby.

References

Anderson, D. L. (1967). Phase changes in the upper mantle. *Science* **157**, 1165.

Anderson, D. L. (1972). Internal constitution of Mars. *J. geophys. Res.* **77**, 789.

Anderson, D. L. and Kovach, R. L. (1967). The composition of the terrestrial planets. *Earth Planet. Sci. Letters* **3**, 19.

Behannon, K. W. (1968). Intrinsic magnetic properties of the lunar body. *J. geophys. Res.* **73**, 7257.

Binder, A. B. (1966). Mariner IV: Analysis of preliminary photographs. *Science* **152**, 1053.

Binder, A. B. (1969). Internal structure of Mars. *J. geophys. Res.* **74**, 3110.

Binder, A. B. and Davis, D. R. (1973). Internal structure of Mars. *Phys. Earth Planet. Int.* **7**, 477.

Brett, R. (1973). A lunar core of Fe–Ni–S. *Geochim. cosmochim. Acta* **37**, 165.

Brett, R. and Bell, P. M. (1969). Melting relations in the Fe-rich portion of the system Fe–FeS at 30 kb pressure. *Earth Planet. Sci. Letters* **6**, 479.

Bridge, H. S., Lazarus, A. J., Snyder, C. W., Smith, E. J., Davis, L. Jr., Coleman P. J., Jr. and Jones, D. E. (1967). Mariner 5, Plasma and magnetic fields observed near Venus. *Science* **158**, 1669.

Bullen, K. E. (1973). On planetary cores. *The Moon* **7**, 384.

Cain, D. L., Kliore, A. J., Seidel, B. L. and Sykes, M. L. (1972). The shape of Mars from the Mariner 9 occultations. *Icarus* **17**, 517.

Carpenter, R. L. (1964). Study of Venus by CW radar. *Astron. J.* **69**, 2.

Carpenter, R. L. (1970). A radar determination of the rotation of Venus. *Astron. J.* **75**, 61.

Chapman, C. R. (1968). Optical evidence on the rotation of Mercury. *Earth Planet. Sci. Letters* **3**, 351.

Colburn, D. S., Currie, R. G., Michalov, J. D. and Sonett, C. P. (1967). Diamagnetic solar wind cavity discovered behind the moon. *Science* **158**, 1040.

Coleman, P. J., Lichtenstein, B. R., Russell, C. T., Sharp, L. R. and Schubert, G. (1972). Magnetic fields near the moon. *Geochim. cosmochim. Acta* **36**, Suppl. 3, 2271.

Colombo, G. (1965). Rotational period of the planet Mercury. *Nature* **208**, 575.

Colombo, G. and Shapiro, I. (1966). The rotation of the planet Mercury. *Astrophys. J.* **145**, 296.

Dolginov, Sh. Sh., Yeroshenko, Ye. G., Zhuzgov, L. N. and Pushkov, N. V. (1966). Measurements of the magnetic field in the vicinity of the moon by the artificial satellite Luna 10. *Dokl. Akad. Nauk SSSR* **170**, 574.

Dolginov, Sh. Sh., Yeroshenko, Ye. G. and Zhuzgov, L. N. (1968). *Kosmich Issled* **6**, 651.

Dolginov, Sh. Sh., Yeroshenko, Ye. G. and Davis, L. (1969). *Kosmich Issled* **7**, 747.

Dolginov, Sh. Sh., Yeroshenko, Ye. G. and Zhuzgov, L. N. (1972). *Dokl. Akad. Nauk SSSR* **207**, 1296.

Dolginov, Sh. Sh., Yeroshenko, Ye. G. and Zhuzgov, L. N. (1973). Magnetic field in the very close neighbourhood of Mars according to data from the Mars 2 and Mars 3 space craft. *J. geophys. Res.* **78**, 4779.

Fjeldbo, G., Fjeldbo, W. C. and von Eshleman, R. (1966). Atmosphere of Mars: Mariner IV models compared. *Science* **153**, 1518.

Frank, L. A., van Allen, J. A. and Hills, H. K. (1963). Mariner 2: preliminary report on measurements of Venus-Charged particles. *Science* **139**, 905.

Fuller, M. (1974). Lunar magnetism. *Rev. Geophys. Space Phys.* **12**, 23.

Goldreich, P. and Peale, S. J. (1966). Resonant spin states in the solar system. *Nature* **209**, 1078.

Goldreich, P. and Peale, S. J. (1967). Spin-orbit coupling in the solar system. II: The resonant rotation of Venus. *Astron. J.* **72**, 662.

Goldreich, P. and Peale, S. J. (1970). The obliquity of Venus. *Astron. J.* **75**, 273.

Goldstein, R. M. (1964). Venus characteristics by Earth-based radar. *Astron. J.* **69**, 12.

Gose, W. A., Peace, G. W., Strangway, D. W. and Larson, E. E. (1972). On the applicability of lunar breccias for paleo-magnetic interpretation. *The Moon* **5**, 106.

Gringauz, K. I., Bezrukikh, V. V., Musatov, L. S. and Breus, T. K. (1968). *Kosmich Issled* **6**, 411.

Gringauz, K. I., Bezrukikh, V. V., Volkov, G. I., Musatov, L. S. and Breus, T. K. (1970). *Kosmich Issled* **8**, 431.

Hall, H. T. and Murthy, V. R. (1971). The early chemical history of the Earth: some critical elemental fractionations. *Earth Planet. Sci. Letters* **11**, 239.

Hanks, T. C. and Anderson, D. L. (1969). The early thermal history of the Earth. *Phys. Earth Planet. Int.* **2**, 19.

Hashin, Z. and Shtrikman, S. (1963). A variational approach to the theory of the elastic behaviour of multiphase materials. *J. Mech. Phys. Solids* **11**, 127.

Johnston, D. H., McGetchin, T. R. and Toksöz, M. N. (1974). The thermal state and internal structure of Mars. *J. geophys. Res.* **79**, 3959.

Levy, E. H. (1972). Magnetic dynamo in the moon: a comparison with the Earth. *Science* **178**, 52.

Lewis, J. S. (1971). Consequences of the presence of sulphur in the core of the Earth. *Earth Planet. Sci. Letters* **11**, 130.

Lewis, J. S. (1972). Metal/silicate fractionation in the solar system. *Earth Planet. Sci. Letters* **15**, 286.

Lyon, E. F., Bridge, H. S. and Binsack, J. H. (1967). Explorer 35 plasma measurements in the vicinity of the moon. *J. geophys. Res.* **72**, 6113.

MacDonald, G. J. F. (1962). On the internal constitution of the inner planets. *J. geophys. Res.* **67**, 2945.

Mariner Stanford Group (1967). Venus: ionosphere and atmosphere as measured by dual frequency radio occultation of Mariner 5. *Science* **158**, 1678.

McConnell, R. K., Jr. and Gast, P. W. (1972). Lunar thermal history revisited. *The Moon* **5**, 41.

Nakamura, Y., Lammlein, D., Latham, G., Ewing, M., Dorman, J., Press, F. and Toksöz, N. (1973). New seismic data on the state of the deep lunar interior. *Science* **181**, 49.

Ness, N. F., Behannon, K. W., Scearce, C. S. and Cantarano, S. C. (1967). Early results from the magnetic field experiment on Lunar Explorer 35. *J. geophys. Res.* **72**, 5769.

Ness, N. F., Behannon, K. W., Lepping, R. P., Whang, Y. C. and Schatten, K. H. (1974a). Magnetic field observations near Venus: preliminary results from Mariner 10. *Science* **183**, 1301.

Ness, N. F., Behannon, K. W., Lepping, R. P., Whang, Y. C. and Schatten, K. H. (1974b). Magnetic field observations near Mercury: preliminary results from Mariner 10. *Science* **185**, 151.

O'Gallagher, J. J. and Simpson, J. A. (1965). Search for trapped electrons and a magnetic moment at Mars by Mariner IV. *Science* **149**, 1233.

Pettengill, G. H. and Dyce, R. B. (1965). A radar determination of the rotation of the planet Mercury. *Nature* **206**, 1240.

Plagemann, S. (1965). A model of the internal constitution and temperature of the planet Mercury. *J. geophys. Res.* 70, 985.

Ringwood, A. E. (1959). On the chemical evolution and densities of the planets. *Geochim. cosmochim. Acta* 15, 257.

Ringwood, A. E. (1966). Chemical evolution of the terrestrial planets. *Geochim. cosmochim. Acta* 30, 41.

Ringwood, A. E. and Essene, E. (1970). Petrogenesis of Apollo 11 basalts, internal constitution, and origin of the moon. *Geochim. cosmochim. Acta* Suppl. 1, 769.

Ringwood, A. E. and Clark, S. P. (1971). Internal constitution of Mars. *Nature* 234, 89.

Runcorn, S. K. (1972). Implications of the magnetism and figure of the moon, Lunar Science III, 666, Lunar Sci. Inst. Contr. No. 88, 1972.

Runcorn, S. K., Collinson, D. W., O'Reilley, W., Battey, M. H., Stephenson, A., Jones, J. M., Manson, A. J. and Readman, P. W. (1970). Magnetic properties of Apollo 11 lunar samples. *Geochim. cosmochim. Acta* 34, 2369.

Runcorn, S. K., Collinson, D. W., O'Reilley, W., Stephenson, A., Battey, M. H., Manson, A. J. and Readman, P. W. (1971). Magnetic properties of Apollo 12 lunar samples. *Proc. R. Soc.* A325, 1571.

Siegfried, R. W. II and Solomon, S. C. (1974). Mercury: internal structure and thermal evolution. *Icarus* 23, 192.

Smith, B. A. and Rees, E. J. (1968). Mercury's rotation period—photographic confirmation. *Science* 162, 1275.

Smith, E. J., Davies, L., Jr., Coleman, P. J., Jr. and Sonett, C. P. (1963). Mariner 2: preliminary report on measurements of Venus: magnetic field. *Science* 139, 909.

Smith, E. J., Davies, L., Jr., Coleman, P. J., Jr. and Jones, D. E. (1965a). Magnetic field measurements near Mars. *Science* 149, 1241.

Smith, E. J., Davies, L., Jr., Coleman, P. J., Jr. and Sonett, C. P. (1965b). Magnetic measurements near Venus. *J. geophys. Res.* 70, 1571.

Solomon, S. C. and Toksöz, M. N. (1973). Internal constitution and evolution of the moon. *Phys. Earth Planet. Int.* 7, 15.

Sonett, C. P., Colburn, D. S. and Currie, R. G. (1967). The intrinsic magnetic field of the moon. *J. geophys. Res.* 72, 5503.

Sonett, C. P., Schubert, G., Smith, R. K., Schwartz, K. and Colburn, D. S. (1971). Lunar electrical conductivity from Apollo 12 magnetic measurements: compositional and thermal inferences. *Geochim. cosmochim. Acta* 35, Suppl. 2, 2415.

Sterrett, K. F., Klement, W. and Kennedy, G. C. (1965). Effect of pressure on the melting of iron. *J. geophys. Res.* 70, 1979.

Strangway, D. W., Pearce, G. W., Gose, W. A. and Timme, R. W. (1971). Remanent magnetization of lunar samples. *Earth Planet. Sci. Letters* 13, 43.

Strangway, D. W. and Sharpe, H. A. (1974). Lunar magnetism and an early cold moon. *Nature* 249, 227.

Toksöz, M. N., Solomon, S. C., Minear, J. W. and Johnston, D. H. (1972). Thermal evolution of the moon. *The Moon* 4, 190.

Tozer, D. C. (1972). The Moon's thermal state and an interpretation of the lunar electrical conductivity distribution. *The Moon* 5, 90.

Usselman, T. M. (1972). The Fe–FeS system at high pressures and the chemical zonation of the core. Int. Conf. Core-Mantle Interface, *Trans. Am. geophys. Un.* **53**, 603.

Van Allen, J. A., Frank, L. A., Krimigis, S. M. and Hills, H. K. (1965). Absence of Martian radiation belts and implications thereof. *Science* **149**, 1228.

Van Allen, J. A., Krimigis, S. M., Frank, L. A. and Armstrong, T. M. (1967). Venus: an upper limit on intrinsic magnetic dipole moment based on absence of a radiation belt. *Science* **158**, 1673.

Wasserburg, G. J., MacDonald, G. J. F., Hoyle, F. and Fowler, W. A. (1964). Relative contribution of uranium, thorium and potassium to heat production in the Earth. *Science* **143**, 465.

Wood, J. A. (1972). Thermal history and early magnetism in the moon. *Icarus* **16**, 229.

Young, R. E. and Schubert, G. (1974). Temperatures inside Mars: is the core liquid or solid? *Geophys. Res. Letters* **1**, 157.

Appendix A

Toroidal and Poloidal Vector Fields

Vector fields T and P are said to be respectively toroidal and poloidal if they can be written

$$T = \operatorname{curl} \frac{\Psi}{r} r = \operatorname{grad} \frac{\Psi}{r} \times r \tag{1}$$

$$P = \operatorname{curl} \operatorname{curl} \frac{\Phi}{r} r = \operatorname{curl} \operatorname{grad} \frac{\Phi}{r} \times r \tag{2}$$

where Ψ and Φ are arbitrary scalar functions of position. It is clear that both T and P are solenoidal and that any solenoidal vector Q may be expressed as a sum of a toroidal and a poloidal vector. In spherical polar coordinates, the components of T and P are

$$T_r = 0, \quad T_\theta = \frac{1}{r \sin\theta} \frac{\partial \Psi}{\partial \phi}, \quad T_\phi = \frac{-1}{r} \frac{\partial \Psi}{\partial \theta} \tag{3}$$

$$P_r = \frac{1}{r^2} L\Phi, \quad P_\theta = \frac{1}{r} \frac{\partial^2 \Phi}{\partial r \partial \theta}, \quad P_\phi = \frac{1}{r \sin\theta} \frac{\partial^2 \Phi}{\partial r \partial \phi} \tag{4}$$

where L^2 stands for the operator,

$$L^2 = \frac{-1}{\sin\theta} \frac{\partial}{\partial \theta} \sin\theta \frac{\partial}{\partial \theta} - \frac{1}{\sin^2\theta} \frac{\partial^2}{\partial \phi^2}. \tag{5}$$

If we have rotational symmetry, $T_r = T_\theta = 0$ and $P_\phi = 0$. If these vectors represent magnetic fields, the field lines for a toroidal field are circles about the axis, while for a poloidal field, the field lines lie in meridional planes.

There are a number of orthogonality properties of the fields P and T for integrations over a sphere of radius r (see e.g. Chandrasekhar, 1961). In particular every poloidal field is orthogonal to every toroidal field. In geophysical problems, it is usual to write the generating functions Ψ and Φ in terms of spherical harmonics $Y_n^m(\theta, \phi) = P_n^m(\cos\theta)e^{im\phi}$ where $P_n^m(\cos\theta)$ are the associated Legendre polynomials. We can thus write

$$\left. \begin{array}{l} \Psi_n^{ms} = R(r)\, P_n^m(\cos\theta)\, \sin m\phi \\[2mm] \Psi_n^{mc} = R(r)\, P_n^m(\cos\theta)\, \cos m\phi \end{array} \right\} \tag{6}$$

where $R(r)$ is a function of r only.

Reference

Chandrasekhar, S. (1961). Hydrodynamic and Hydromagnetic Stability. Clarendon Press, Oxford.

Author Index

Numbers in italic indicate those pages where references are given in full

C

D

Subject Index